Lecture Notes in Computer Science 5104

Commenced Publication in 1973
Founding and Former Series Editors:
Gerhard Goos, Juris Hartmanis, and Jan van Leeuwen

Fernando Bello
P.J. "Eddie" Edwards (Eds.)

Biomedical Simulation

4th International Symposium, ISBMS 2008
London, UK, July 7–8, 2008
Proceedings

 Springer

Volume Editors

Fernando Bello
P.J. "Eddie" Edwards
Department of Biosurgery and Surgical Technology
Imperial College London, St. Marys Hospital Campus
London W2 1NY, United Kingdom
E-mail:{f.bello; eddie.edwards}@imperial.ac.uk

Library of Congress Control Number: Applied for

CR Subject Classification (1998): I.6, I.4, J.3, I.3

LNCS Sublibrary: SL 1 – Theoretical Computer Science and General Issues

ISSN 0302-9743
ISBN 978-3-540-70520-8 Springer Berlin Heidelberg New York

Springer is a part of Springer Science+Business Media

springer.com

© Springer-Verlag Berlin Heidelberg 2008

Typesetting: Camera-ready by author, data conversion by Scientific Publishing Services, Chennai, India
Printed on acid-free paper   SPIN: 12324446   06/3180   5 4 3 2 1 0

Preface

This book contains the written contributions to the 4th International Symposium on Biomedical Simulation (ISBMS), which was held at Imperial College in London, UK, July 7–8, 2008.

The manuscripts are organized in four different sections corresponding to key areas and techniques of this constantly expanding field: finite element modeling, mass spring and statistical shape modeling, motion and fluid modeling and implementation issues. An additional section covers the posters presented at the meeting.

Recent years have seen increased interest in computational models for biomedical simulation. The symposium provided an international forum for researchers in this area to share their latest work and to discuss future trends. It was held in the spirit and continuation of the symposia on Surgical Simulation and Soft Tissue Modeling (IS4TM) organized in 2003 by INRIA, Medical Simulation (ISMS) in 2004 by CIMIT and Biomedical Simulation (ISBMS) in 2006 by ETH.

The meeting was single track with 19 oral presentations and 7 posters selected after evaluation by at least three members of the Program Committee. In addition to contributed papers, the symposium included discussion panels on meshless methods and GPU-based simulation that gave participants an opportunity to learn the principles, applications, implementation details, and discuss future developments in these increasingly relevant areas. Denis Noble and Guang-Zhong Yang gave keynote talks on their work in computational modeling of biological systems and image-constrained biomechanical modeling, respectively.

The geographical breakdown of the different institutions presenting their research was: Australia, Canada, Denmark, France, Germany, Spain, Switzerland and UK. The quality and breadth of the contributions indicate that the symposium continues to be an important forum for our rapidly growing field, bringing together several related disciplines.

We would like to thank the Program Committee members for volunteering their time to review and discuss symposium papers and doing so in a timely and professional manner. We are also thankful to the Steering Committee for their encouragement and support to continue the tradition laid by the previous meetings in Juan-Les-Pins, Cambridge MA and Zurich. Special thanks go to the Local Organizing Committee for their hard work in making ISBMS 2008 a successful event. Last but not least, we would like to thank all authors for presenting their work at the symposium. It was a pleasure hosting ISBMS 2008 and we hope that all participants enjoyed the intense and stimulating symposium.

July 2008

Fernando Bello
P.J. "Eddie" Edwards

Organization

The Fourth International Symposium on Biomedical Simulation ISBMS 2008 was organized by the departments of Biosurgery and Surgical Technology and Computing of Imperial College London.

ISBMS 2008 would not have been possible without the dedication of the Local Organizing Committee: Harry Brenton, Askanksha Chhikara, Vincent Luboz, and Pierre-Frédéric Villard.

Steering Committee

Fernando Bello, Chair	Imperial College London, UK
P.J. "Eddie" Edwards, Co-chair	Imperial College London, UK
Nicholas Ayache	INRIA, France
Stéphane Cotin	INRIA, France
Steve Dawson	CIMIT, USA
Hervé Delingette	INRIA, France
Matthias Harders	ETH Zurich, Switzerland
Dimitris Metaxas	Rutgers University, USA
Sebastien Ourselin	University College London, UK
Gábor Székely	ETH Zurich, Switzerland

Program Committee

Jérémie Allard	INRIA, France
Remis Balaniuk	CU Brasilia, Brazil
Cagatay Basdogan	KOC University, Turkey
Eliías Cueto	Zaragoza University, Spain
Christos Davatzikos	University of Pennsylvania, USA
James Duncan	Yale University, USA
Alejandro Frangi	Pompeu Fabra University, Spain
Anthony G. Gallagher	RCSI, Ireland
Miguel A. González Ballester	Bern University, Switzerland
Vincent Hayward	McGill, Canada
Karl Heinz Höhne	University Hospital Hamburg, Germany
Arun Holden	Leeds University, UK
David Holmes	Mayo Clinic, USA
Peter Hunter	University of Auckland
Edoardo Mazza	ETH Zurich, Switzerlad
Vincent Luboz	Imperial College London, UK
Michael Miga	Vanderbilt University, USA

Karol Miller	University of Western Australia
Paul Neuman	CIMIT, USA
Wiro Niessen	Erasmus MC Rotterdam, The Netherlands
Mark Ottensmeyer	CIMIT, USA
Dinesh Pai	Rutgers University, USA
Luc Soler	IRCAD, France
T.S. Sørensen	Århus University, Denmark
Dominik Szczerba	ETH Zurich, Switzerland
Frank Tendick	UC San Francisco, USA
Daniel Thalman	EPFL, Switzerland
Pierre-Frédéric Villard	Imperial College London, UK
Simon Warfield	Harvard and Childrens Hospital, USA
Xunlei Wu	Renci, USA
Guang-Zhong Yang	Imperial College London, UK

Table of Contents

Finite Element Modelling

Mass Spring and Statistical Shape Modelling

Motion and Fluid Modelling

Implementation Issues

Posters

Physically Based Finite Element Model of the Face

Giuseppe Barbarino[1], Mahmood Jabareen[1], Juergen Trzewik[2], and Edoardo Mazza[1,3]

[1] Institute of Mechanical Systems,
Department of Mechanical and Process Engineering,
ETH Zurich,
8092 Zurich, Switzerland
Edoardo.Mazza@imes.mavt.ethz.ch
[2] Johnson & Johnson MEDICAL GmbH,
22851 Norderstedt, Germany
[3] EMPA, Swiss Federal Laboratory for Materials Testing & Research
8600 Dübendorf, Switzerland

Abstract. The proposed 3D finite element model of the face aims at a faithful representation of the anatomy, the mechanical interactions between different tissues, and the non linear force deformation characteristics of tissues. Bones and soft tissues were reconstructed from magnetic resonance images. Non linear constitutive equations are implemented in the numerical model. The corresponding model parameters were selected according to previous work with mechanical measurements on soft facial tissue. Model assumptions concerning tissues geometry, mechanical properties and boundary conditions were validated through comparison with measurements of the facial tissue response to gravity loads and to the application of a pressure inside the oral cavity. In particular, parametric studies were carried out in order to quantify the influence of constitutive model parameters of muscles. The model described in this paper might be used for simulation of plastic and reconstructive surgery and for investigation of the physiology and pathology of face deformation.

Keywords: Face, numerical modeling, validation, constitutive equations, muscles.

1 Introduction

The deformation behavior of soft facial tissue has been investigated in several studies aiming at the realization of realistic animations of facial expressions, see e.g. [1]. Simulations were proposed related to the physiology and pathology of face deformation, developing finite element models of human mastication [2], speech [3], for planning craniofacial [4] and maxillofacial [5, 6] surgery. For clinical applications, challenges are related to the reliability of the models with respect to the anatomy, the mechanical interactions between different tissues, and the non linear (time dependent) force deformation characteristics of soft biological materials. Face models were proposed with uniform properties for all soft tissues, Keeve [7], or distinct constitutive equations for different tissue layers and explicit representation of face muscles, Chabanas [8]. In [4] and [9] 18 face muscles were modeled based on segmentation of MRI images and activated through a force field in the tangent direction of the fibers.

F. Bello, E. Edwards (Eds.): ISBMS 2008, LNCS 5104, pp. 1–10, 2008.

Most numerical models proposed so far utilize linear elastic material equations for soft facial tissue. Non linear constitutive models are required in case of large tissue deformations, [10]. As an example, in [11] hyperelastic-viscoplastic constitutive equations with internal variables (including a so called "aging function") were used to simulate the progressive loss of stiffness of the aging facial tissue.

The 3D finite element model of the face presented in this work is part of a common project of ETH Zurich and Johnson & Johnson MEDICAL GmbH aiming at improving the design of medical devices (such as suture products) used for plastic surgery on the human face. Bones, muscles, skin, fat, and superficial muscoloaponeurotic system were reconstructed from magnetic resonance images and modeled according to anatomical, plastic and reconstructive surgery literature. Organs and tissue geometry, constraints (ligaments) and interaction of tissues (contact interfaces) were verified to be consistent with state of the art knowledge on face anatomy. Non linear constitutive equations were implemented into the numerical algorithms for modeling the mechanical behavior of each tissue. The model generation procedure, first calculations and the results of face aging predictions were described in [12]. The main uncertainty in these simulations is related to the constitutive equations utilized to represent the deformation behavior of soft face tissue. In our model, formulation of constitutive equations and the corresponding materials parameters are selected based on previous work with mechanical measurements on some facial tissue, [11, 13-19]. The scatter of data proposed in literature is particularly large for skeletal muscles, with stiffness parameters differing by two orders of magnitude depending on the experimental conditions and data analysis procedure applied. Calculations are presented in this paper with different set of material parameters applied to determine the response of facial tissues to gravity loads and to the application of a pressure inside the oral cavity. Measurements were performed for these loading cases using magnetic resonance images and holographic techniques. Comparison of calculated and measured face deformation provides a validation for the proposed model and shows the influence of the soft tissue constitutive model on the results of face simulations.

2   Finite Element Model

2.1 Mesh, Interactions and Kinematic Boundary Conditions

The model presented was based on magnetic resonance imaging (MRI) data. Data were acquired with a Philips Achieva 1.5T scanner. The entire head and neck of a 27 years old male lying in the supine position were scanned. A field of view of 200x200x300 mm, an in-plane resolution of 0.5 mm and a slice distance of 2 mm have been set, for a total of 150 slices. Fig. 1 shows transversal MR images of the face in different heights and examples of transversal, coronal and sagittal sections.

MR images provide detailed information about internal structures, tissues and organs. Segmentation and reconstruction of each single anatomical part in form of triangulated surfaces and creation of a polygonal surface model was achieved using the software AMIRA (http://www.amiravis.com/). This model in form of triangulated surfaces was imported into the software Raindrop Geomagic (http://www.geomagic.com/en/products/studio/), where single parts were smoothed and incomplete objects corrected, in line with descriptions from anatomical literature, see [12].

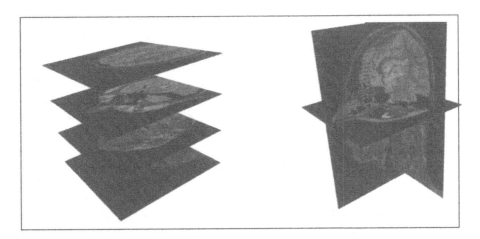

Fig. 1. Examples of MR scan image set. Left: Transversal sections. Right: sagittal, coronal and transverse section.

Skull and Mandibula have been reconstructed from the MR scans with a straight and quasi-automatic procedure. They constitute the most internal part of the model and provide anchoring points for the muscles, the mucosa and fixation points for the whole model. The most relevant muscles have been considered in the model (they are indicated Fig. 2). Retaining ligaments (which couldn't be identified from the MR images) are modeled through application of kinematic constraints to the corresponding nodes of the finite element model, see [12]. Ligament location was selected according to anatomic literature. Skin (epidermis and dermis) is modeled with a layer of constant thickness (2 mm) all over the face. The assumption of uniform skin thickness was validated through measurement performed on MR slides. It was found that in the region of interest the thickness variation is rather small. Mucosa is the innermost layer, which is continuous and similar in structure to the skin, thus it was modeled with the same thickness and mechanical properties. The SMAS (Superficial Muscolo Aponeurotic System) is a non-uniform layer of connective tissue consisting of fat cells, collagen, elastic and muscle fibers. The surface dividing SMAS and deeper connective tissue cannot be clearly identified in the MR images. The model assumption adopted here is to introduce the SMAS together with the superficial fat as a layer parallel to the skin with a thickness of 3 mm. Connective tissue deeper to the SMAS mainly consists of fat cells; it was introduced as filler of the empty spaces. As to the method adopted for tissue and organ reconstruction, we proceeded as follows: skin and SMAS were reconstructed first as the two outermost layers, then the mucosa, the innermost layer. SMAS and skin share a common surface; mucosa is continuous with the skin in the lips region. Muscles have been then reconstructed and at locations where they were superposing the SMAS part of this layer was removed. The remaining empty spaces in between Smas, Mucosa, muscles and bones were filled with connective tissue.

All parts have been meshed with a semiautomatic procedure which can generate hexahedron meshes starting from an object in form of a triangulated surface. This procedure is described in [12]. Each tissue was meshed separately, except for skin and

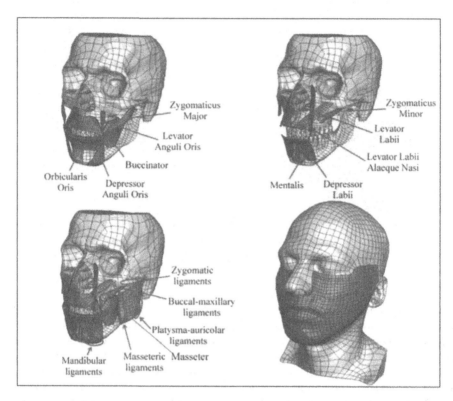

Fig. 2. FE face model: muscles and ligaments are indicated. FE mesh is darker, skull and part of the head reconstructed for display purposes in light grey.

SMAS which share nodes at their interface. All contact interfaces are modeled as tied (no sliding, no separation). The whole model was evaluated and approved by an anatomist. The final finite element model shown in Fig. 2 consists of 3874 hexahedron elements and 8395 nodes.

2.2 Constitutive Model

Constitutive model formulation is based on the work of Rubin and Bodner [13]. The R-B model was implemented as a user defined algorithm into the commercial finite element package ABAQUS (http://www.simulia.com/). Steady response of facial tissue was considered in the experiments and simulations performed for model validation. For this reason the parts of the model that describe the transient and dissipative material behavior are neglected and only the hyperelastic terms are active. The elastic part of the R-B model is briefly introduced here.

The specific (per unit mass) strain energy function Ψ is given by

$$\rho_0 \Psi = \frac{\mu_0}{2q} \left[e^{qg} - 1 \right] \tag{1}$$

where ρ_0 is the mass density in the reference configuration, and μ_0 and q are material parameters. The function g is specified in an additive form

$$g = g_1 + g_2 = 2m_1[J - 1 - \ln(J)] + m_2(\beta_1 - 3) \tag{2}$$

with g_1 and g_2 characterizing the response of the fully elastic dilatation, and fully elastic distortion, respectively. It should be noted that the R-B model has the capability to characterize the response of fibers as well as the visco-plasticity associated with dissipative components of the material response. In the present paper these effects are not considered. In equation (2) J is a measure of dilatation, and β_1 is a pure measure of the elastic distortion, given by

$$J = det(\mathbf{F}) \tag{3}$$

$$\beta_1 = \bar{\mathbf{B}}{:}\mathbf{I} \tag{4}$$

where $\mathbf{F}$ is the deformation gradient, and $\bar{\mathbf{B}} = J^{-2/3}\mathbf{B} = J^{-2/3}\mathbf{F}\mathbf{F}^T$ is the modified left Cauchy-Green deformation tensor. Finally, the corresponding Cauchy stress tensor is given by

$$\boldsymbol{\sigma} = m_1\mu\left[1 - \frac{1}{J}\right]\mathbf{I} + m_2\mu\bar{\mathbf{B}}' \tag{5}$$

with $\bar{\mathbf{B}}'$ (the deviatoric part of the modified left Cauchy-Green deformation tensor) and μ given as:

$$\mu = \mu_0 e^{qg} \tag{6}$$

$$\bar{\mathbf{B}}' = \bar{\mathbf{B}} - \frac{1}{3}(\bar{\mathbf{B}}{:}\mathbf{I})\mathbf{I} \tag{7}$$

For small deformations the parameters $m_1{\cdot}\mu$ and $m_2{\cdot}\mu$ (see equation 5) correspond to bulk modulus and shear modulus respectively. Five material parameters have to be assigned to the different facial tissues. They are reported in Table 1 for each tissue. The parameters were determined from values reported in literature for skin, mucosa and fat, [11, 13, 14]. Mechanical properties of muscles are defined as three different sets (set1, set2, set3), with elastic constants (in particular, m_2, associated with shear

Table 1. Material parameters for the R-B model

	ρ (g/cm^3)	μ_0 (MPa)	q	m_1	m_2
Skin/mucosa	1.1	1.7	36	1294	0.0008
SMAS	1	3.7	25	595	0.0008
Deep fat	1	3.7	25	595	0.00007
Muscles-set1	1	3.7	25	595	0.00009
Muscles-set2	1	3.7	25	595	0.0022
Muscles-set3	1	3.7	25	595	0.009

modulus) ranging over two orders of magnitude. This reflects the scatter of the values reported in literature for skeletal muscles, see [15-19]. The significance of this uncertainty for the deformation behavior of the face is demonstrated in section 3. Note that the muscles are represented here with an isotropic constitutive model, thus neglecting anisotropy arising from muscle fiber orientation.

3   Model Validation and Discussion

The numerical model requires validation due to the uncertainties related to (i) constitutive model parameters, (ii) influence of spatial discretization (mesh density), (iii) boundary conditions (ligaments, fixations) and organ interactions. For this purpose, the response of facial tissues in terms of face shape changes consequent to the application of mechanical loads was analyzed.

MRI scans were acquired for the "reference" configuration (from which the face model was generated, with the tested person lying on the nape in the supine position facing the ceiling), and for a "deformed" configuration with the tested person lying horizontally facing the floor (in this position forehead and chest lye on rigid support, without interference with the skin surface of cheeks and neck). The differences in the shape of external contour of these two measurements allow evaluating the effect of gravity. Specifically, the differences have been quantified by computing the shortest distance between the skin surfaces. The same procedure has been used to evaluate skin surface displacements calculated with the finite element model. The gravity field is considered to have positive sign in the direction from the nape towards the nose. In the "reference" configuration a gravity field of -g is acting on the body, while during the second measurement the acting gravity field is +g. Thus, a value of +2g was used for the magnitude of the distributed force field in the simulation. This approach obviously represents an approximation of the real load case. This approximation is unavoidable, since any geometry acquisition will be influenced by a gravity field, so that the definition of a reference "undeformed and unloaded" configuration is not possible. The influence of this uncertainty however is expected to be negligible due to the relatively small displacements associated with these loading cases. Results of measurements and calculations (for the different muscle parameters) are presented in Fig. 3.

Two holographic measurements have been performed using the camera set-up described in [20]. The first with the person standing vertically in front of the holographic system with no face muscle contractions, while the second in the same position, but with a tube inserted between the lips and blowing into it with a relative pressure of approximately 10 mbar (measured with a pressure sensor). The face surface displacement is quantified based on the measured distance between the two surfaces, computed from the data sets using same procedure as for the MRI measurements. Results of measurement and corresponding calculations (for the different muscle parameters) are shown in Fig. 4. The displacement distribution in the calculation agrees to a great extent with the measured one for muscle parameter set 2.

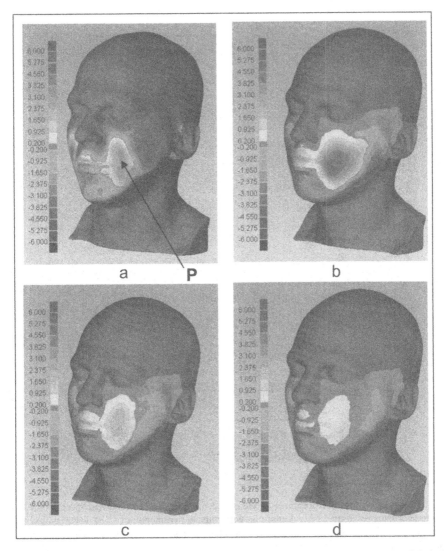

Fig. 3. Response to gravity load: a) Measured relative displacement (in mm). Point P is indicated for later reference. FE calculations with muscle parameter set 1 (b), set 2 (c) and set 3 (d).

The magnitude of displacement of points P and Q (see Figures 3a and 4a) is reported for measurement and calculations in Table 2. Calculations with set1 lead to large overestimation of displacement; with set3 the response is much too stiff. For both load cases closer agreement between measurement and calculation is found for muscle parameter set 2. For this calculation, the region of maximum displacement from gravity loading (Fig. 3) has the same extension and the magnitude of the displacement is in the same order as for the measurement. The same is observed from a comparison of calculated and measured displacement for the load case with internal

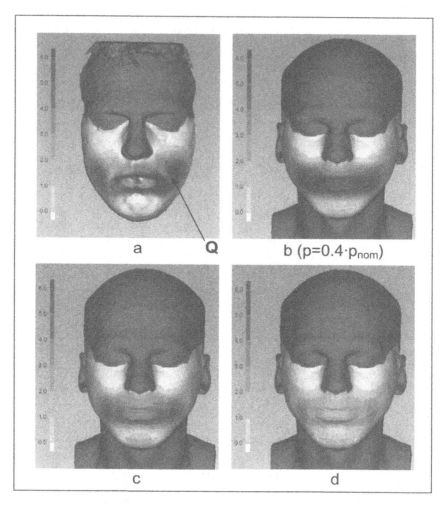

Fig. 4. Response to internal pressure: a) Measured relative displacement (in mm). Point Q is indicated for later reference. FE calculations with muscle parameter set 1 (b), set 2 (c) and set 3 (d). Note, for b: maximum internal pressure: 4 mbar (excessive deformation with convergence failure above this value).

pressure in the oral cavity. Differences in the lips region near the plane of symmetry of the face are due to the presence of the tube connected to a pressure sensor. Lips are modeled as tight together in the FE simulations.

General agreement between measurement and calculation for these two cases is regarded as a confirmation of the predictive capabilities of the present model. The influence of the constitutive model parameters of facial muscles is demonstrated, with significant changes in the magnitude and distribution of the relative displacement calculated with different values of the shear stiffness parameter m_2, which controls the resistance to distortional deformations.

Table 2. Comparison of measured and calculated displacement at selected points

	Measured	Set1	Set2	Set3
Gravity Displacement of point P (mm)	2.8 (left) 2.43 (right)	4.61	2.06	0.64
Internal pressure Displacement of point Q (mm)	5 (left) 4.2 (right)	5.3*	4.12	1.89

*: this value was obtained with 40% of the nominal pressure value, the calculation failed to converge for high values of pressure.

4 Conclusions

A detailed finite element model of the human face has been created based on segmentation of MRI data. Mechanical properties of facial tissues were assigned using data previously proposed in literature and the influence of these parameters on the mechanical behavior has been investigated. The reliability of the proposed model has been verified through comparison with measurements of the response to gravity loads and to the application of a pressure inside the oral cavity.

Future work will consider the representation of both non linear elastic and viscoplastic material behavior. Consideration will be given to the anisotropic behavior of muscles. Further mechanical tests on facial soft tissues will be performed in order to determine suitable sets of material parameters and to verify the influence of model assumptions concerning tissue interactions.

The present model might be used for prediction of the outcome of plastic and reconstructive surgery. For future patient specific analyses, improvements of the model generation procedure with automatic organs and tissue reconstruction, finite element meshing, and mechanical characterization will be essential: consider that the creation of the present FE model from MR images required several weeks of work. Realistic representation of muscle contraction is an important aspect of the model improvements required for simulation of facial expressions or mastication, and this will be addressed in future work. Next to medical studies, the proposed model might also provide benchmark examples for the development of simplified methods for realistic animations of facial expressions.

Acknowledgments

This work was supported by the CO-ME/NCCR research network of the Swiss National Science Foundation.

References

1. Zhang, Y., Prakash, E.C., Sung, E.: Face alive. J. Visual Lang. Comput. 15(2), 125–160 (2004)
2. Rohrle, O., Pullan, A.J.: Three-dimensional finite element modelling of muscle forces during mastication. J. Biomech. 40(15), 3363–3372 (2007)

3. Vogt, F., Lloyd, J.E., Buchaillard, S., Perrier, P., Chabanas, M., Payan, Y., Fels, S.S.: Investigation of Efficient 3D FE Modeling of a Muscle-Activated Tongue. In: Harders, M., Székely, G. (eds.) ISBMS 2006. LNCS, vol. 4072, pp. 19–28. Springer, Heidelberg (2006)

4. Gladilin, E., Zachow, S., Deuflhard, P., Hege, H.C.: Anatomy- and physics-based facial animation for craniofacial surgery simulations. Med. Biol. Eng. Comput. 42(2), 167–170 (2004)

5. Mollemans, W., Schutyser, F., Nadjmi, N., Maes, F., Suetens, P.: Predicting soft tissue deformations for a maxillofacial surgery planning system: From computational strategies to a complete clinical validation. Med. Image Anal. 11(3), 282–301 (2007)

6. Luboz, V., Chabanas, M., Swider, P., Payan, Y.: Orbital and maxillofacial computer aided surgery: patient-specific finite element models to predict surgical outcomes. Comput. Methods Biomech. Biomed. Engin. 8(4), 259–265 (2005)

7. Keeve, E., Girod, S., Kikinis, R., Girod, B.: Deformable modeling of facial tissue for craniofacial surgery simulation. Comput. Aided Surg. 3(5), 228–238 (1998)

8. Chabanas, M., Luboz, V., Payan, Y.: Patient specific finite element model of the face soft tissues for computer-assisted maxillofacial surgery. Med. Image Anal. 7(2), 131–151 (2003)

9. Zachow, S., Gladilin, E., Hege, H.C., Deuflhard, P.: Towards patient specific, anatomy based simulation of facial mimics for surgical nerve rehabilitation. In: CARS 2002: Computer assisted radiology and surgery, Paris, France, pp. 3–6 (2002)

10. Gerard, J.M., Ohayon, J., Luboz, V., Perrier, P., Payan, Y.: Non-linear elastic properties of the lingual and facial tissues assessed by indentation technique - Application to the biomechanics of speech production. Med. Eng. Phys. 27(10), 884–892 (2005)

11. Mazza, E., Papes, O., Rubin, M.B., Bodner, S.R., Binur, N.S.: Nonlinear elastic-viscoplastic constitutive equations for aging facial tissues. Biomech. Model Mechan. 4(2-3), 178–189 (2005)

12. Barbarino G., Jabareen M., Trzewik J., Stamatas G., Mazza E., Development and validation of a 3D finite element model of the face. J Biomech. Eng. (submitted)

13. Rubin, M.B., Bodner, S.R.: A three-dimensional nonlinear model for dissipative response of soft tissue. Int. J. Solids Struct. 39(19), 5081–5099 (2002)

14. Duck, F.A.: Physical properties of tissue, a comprehensive reference book. Academic Press, London (1990)

15. Malinauskas, M., Krouskop, T.A., Barry, P.A.: Noninvasive measurement of the stiffness of tissue in the above knee amputation limb. J. Rehab Res. Dev. 26(3), 45–52 (1989)

16. Dresner, M.A., Rose, G.H., Rossmann, P.J., Multhupillai, R., Manduca, A., Ehman, R.L.: Magnetic resonance elastography of skeletal muscle. J. Magn. Res. Imag. 13, 269–276 (2001)

17. Papazoglou, S., Rump, J.R., Braun, J., Sack, I.: Shear Wave group velocity inversion in MR elastography of human skeletal muscle. J. Magn. Res. Imag. 56, 489–497 (2006)

18. Levinson, S.F., Shinagawa, M., Sato, T.: Sonoelastic determination of human skeletal muscle elasticity. J. Biomech. 28(10), 1145–1154 (1995)

19. Gennisson, J.L., Cornu, Ch., Catheline, S., Fink, M., Portero, P.: Human muscle hardness assessment during incremental isometric contraction using transient elastography. J Biomech 38(7), 1543–1550 (2005)

20. Hirsch, S., Frey, S., Thelen, A., Ladriere, N., Hering, P.: Facial topometry using pulsed holography. In: Proc. 7th Int. Symp. Display Holography, St. Asaph, Wales (2006)

On the Application of Model Reduction Techniques to Real-Time Simulation of Non-linear tissues

Siamak Niroomandi[1], Icíar Alfaro[1], Elías Cueto[1], and Francisco Chinesta[2]

[1] Group of Structural Mechanics and Material Modelling.
Aragón Institute of Engineering Research, I3A. Universidad de Zaragoza.
Edificio Betancourt. María de Luna, 7. E-50012 Zaragoza, Spain
ecueto@unizar.es,
http://gemm.unizar.es
[2] Laboratoire de Mecanique des Systemes et des Procedes.
UMR 8106 CNRS-ENSAM-ESEM. 151, Bvd. de l'Hòpital. 75015 Paris, France
francisco.chinesta@paris.ensam.fr,
http://www.paris.ensam.fr/lmsp/

Abstract. In this paper we introduce a new technique for the real-time simulation of non-linear tissue behavior based on a model reduction technique known as Proper Orthogonal (POD) or Karhunen-Loève Decompositions. The technique is based upon the construction of a complete model (using Finite Element modelling or other numerical technique, for instance, but possibly from experimental data) and the extraction and storage of the relevant information in order to construct a model with very few degrees of freedom, but that takes into account the highly non-linear response of most living tissues. We present its application to the simulation of palpation a human cornea and study the limitations and future needs of the proposed technique.

1 Introduction

Real-time surgery simulation [5] has attracted the attention of a wide community of researchers. The utility of such techniques are obvious, and they include, for instance, surgery planning, training of surgeons in image-guided surgery or minimally-invasive surgery, etc.

The state of the art of the technique has evolved very rapidly, see for instance [8] or [11] for interesting surveys. Starting from spring-mass systems, the nowadays real-time surgical simulators are now mostly based on Finite Element (FE) or Boundary Element (BE) technologies, able to account for a quite realistic behavior and even large deformations, see, for instance, [2][3][6].

Such simulators should provide a physically more or less accurate response such that, with the use of haptic devices, a realistic feedback is transmitted to the surgeon in terms of both visual feedback and force feedback. Following [4], "... the model may be physically correct if it looks right".

F. Bello, E. Edwards (Eds.): ISBMS 2008, LNCS 5104, pp. 11–18, 2008.

For that to be possible, it is commonly accepted that a minimum bandwidth of 20-60 Hz for visual feedback and 300-1000 Hz for haptic display is necessary, see [7]. In this paper we focus our attention in the second requirement for the deformable model. All the simulations performed were designed to run under that requirements.

Very recently, geometric non-linearities have been taken into account in a work also based in model reduction, see [2]. But in this case, only linear materials have been considered (i.e., the so-called Saint Venant-Kirchhoff models, or homogeneous isotropic linear elastic materials undergoing large deformations). Most soft tissues, however, exhibit complex non-linear responses, possibly with anisotropic characteristics, and are frequently incompressible or quasi-incompressible. Geometric non-linearities (those deriving from large strains) should be also taken into consideration on top of this complex material behavior. The correct simulation of these materials requires the employ of Newton-Raphson or similar techniques in an iterative framework. This makes the existing engineering FE codes unpractical for real-time simulations.

The technique here presented is based upon existing data on the behavior of the simulated tissues. These data can be obtained after numerical simulations made off-line and stored in memory. But they can be also obtained from physical experiments, for instance. For the work here presented we have chosen the first option, and FE models of the organs being simulated will be considered as an "exact" to compare with. From these data we extract the relevant information about the (non-linear) behavior of the tissues, with the help of Karhunen-Loève decompositions and employ it to construct a very fast Galerkin method with very few degrees of freedom. To this end, we employ model reduction techniques based on proper orthogonal decompositions [9] [10] [12].

In order to show the performance of the method, we have chosen to simulate the behavior of the human cornea, although the technique is equally applicable to any other soft tissue. The cornea presents a highly non-linear response, with anisotropic and heterogeneous behavior due to its internal collagen fiber reinforcement. As an accurate enough model we have implemented that employed in [1]. This model is briefly reviewed in Section 2.

2   A Hyperelastic Mechanical Model for the Human Cornea

As mentioned before, we have chosen the human cornea as an example of highly non-linear tissue. This non-linearity comes from a variety of reasons, such as the internal collagen fiber reinforcement (material non-linearity) and also from the very large strains it could suffer. The human cornea is composed by a highly porous material, composed by nearly 80% by water, and thus quasi-incompressible. Most of the cornea's thickness (around 90%) constitutes the stroma, that is composed of 300-500 plies of collagen fibers, distributed in parallel to the surface of the cornea. This microstructure induces in the corneal tissue a highly non-linear and heterogeneous behavior.

The model here employed for the simulation of the human cornea [1] considers the cornea as a hyperelastic material. Reinforcing fibers, that move continuously together with the cornea, posses a direction m_0, with $|m_0| = 1$. The fiber stretching after the deformation will be given by $\lambda m(x, t) = Fm_0$, where $F = dx/dX$ represents the deformation gradient. A second family of fibers, n_0, is also considered as reinforcement at each point.

Due to the dependence of strain on the considered direction, the existence of a strain energy density functional, Ψ, depending on the right Cauchy-Green tensor, $C = F^T F$, and the initial fiber orientations, m_0 and n_0, is postulated. Based on the volumetric incompressibility restrictions, this functional can be expressed as [1]

$$\Psi(C) = \Psi_{vol}(J) + \bar{\Psi}(\bar{C}, m_0 \otimes m_0, n_0 \otimes n_0) \tag{1}$$

where $\Psi_{vol}(J)$ describes the volumetric change and $\bar{\Psi}(\bar{C}, m_0 \otimes m_0, n_0 \otimes n_0)$ the change in shape. Both are scalar functions of $J = det F$, $\bar{C} = \bar{F}^T \bar{F}$, where $\bar{F} = J^{-1/3} F$, m_0 and n_0.

Once this energy density functional is known, the second Piola-Kirchhoff stress tensor, S, and the fourth-order tangent constitutive tensor, C, can be determined by

$$S = 2\frac{\partial \Psi}{\partial C} \qquad C = 2\frac{\partial S(C)}{\partial C} \tag{2}$$

A detailed derivation of the model can be obtained in [1]. The interested reader is referred to this paper for reference.

3   Model Reduction Techniques

3.1   Fundamentals: Karhunen-Loève or Proper Orthogonal Decomposition

In Karhunen-Loève techniques [9] we assume that the evolution of a certain field $T(x, t)$ is known. In practical applications (assume that we have performed off-line some numerical simulations, for instance), this field is expressed in a discrete form which is known at the nodes of a spatial mesh and for some times t^m. Thus, we consider that $T(x_i, t^m) = T^m(x_i) \equiv T_i^m$ ($t^m = m \times \Delta t$) are known. We can also write T^m for the vector containing the nodal degrees of freedom at time t^m. The main idea of the Karhunen-Loève (KL) decomposition is to obtain the most typical or characteristic structure $\phi(x)$ among these $T^m(x)$, $\forall m$. This is equivalent to obtain a function that maximizes α:

$$\alpha = \frac{\sum_{m=1}^{m=M} \left[\sum_{i=1}^{i=N} \phi(x_i) T^m(x_i) \right]^2}{\sum_{i=1}^{i=N} (\phi(x_i))^2} \tag{3}$$

where N represents the number of nodes of the complete model and M the number of computed time steps. The maximization leads to an eigenvalue problem of the form:

$$\tilde{\phi}^T c \, \phi = \alpha \tilde{\phi}^T \phi; \; \forall \tilde{\phi} \Rightarrow c \, \phi = \alpha \phi \tag{4}$$

where we have defined the vector ϕ such that its i-th component is $\phi(x_i)$, and the two-point correlation matrix, c, is given by

$$c_{ij} = \sum_{m=1}^{m=M} T^m(x_i) T^m(x_j) \Leftrightarrow c = \sum_{m=1}^{m=M} T^m (T^m)^T \qquad (5)$$

which is symmetric and positive definite. If we define the matrix Q containing the discrete field history:

$$Q = \begin{pmatrix} T_1^1 & T_1^2 & \cdots & T_1^M \\ T_2^1 & T_2^2 & \cdots & T_2^M \\ \vdots & \vdots & \ddots & \vdots \\ T_N^1 & T_N^2 & \cdots & T_N^M \end{pmatrix} \qquad (6)$$

then it is easy to verify that the matrix c in Eq. (4) results in $c = Q \, Q^T$.

3.2   A Posteriori Reduced Modelling of Transient Models

If some direct simulations have been carried out, we can determine T_i^m, $\forall i \in [1, \cdots, N]$ and $\forall m \in [1, \cdots, M]$, and from these solutions the n eigenvectors related to the n-highest eigenvalues that are expected to contain the most important information about the problem solution. For this purpose we solve the eigenvalue problem defined by Eq. (4) retaining all the eigenvalues ϕ_k belonging to the interval defined by the highest eigenvalue and that value divided by a large enough value (10^8 in our simulations). In practice n is much lower than N, and this constitutes the main advantage of the technique. Thus, we can try to use these n eigenfunctions ϕ_k for approximating the solution of a problem *slightly* different to the one that has served to define T_i^m. For this purpose we need to define the matrix $B = [\phi_1 \cdots \phi_n]$.

If we now consider the linear system of equations coming from the discretization of a generic problem, in the form $G \, T^m = H^{m-1}$, where the superscript refers to the time step, then, assuming that the unknown vector contains the nodal degrees of freedom, it can be expressed as a linear combination of eigenmodes:

$$T^m = \sum_{i=1}^{i=n} \zeta_i^m \, \phi_i = B \, \zeta^m \qquad (7)$$

where ζ_i^m represent the new degrees of freedom of the problem, from which we obtain

$$G \, T^m = H^{m-1} \Rightarrow G \, B \, \zeta^m = H^{m-1} \qquad (8)$$

and by multiplying both terms by B^T we obtain $B^T G \, B \, \zeta^m = B^T H^{m-1}$, which proves that the final system of equations is of low order, i.e. the dimension of $B^T G \, B$, is $n \times n$, with $n \ll N$.

4   Numerical Results

In order to test the performance of the proposed technique, we have focused our attention mainly in two aspects. First, the accuracy of the results. Second, the compliance with the requirements of haptic feedback, i.e., all results must be obtained at a frequency between 300 and 1000 Hz.

A set of tests have been accomplished, all based in the model of the human cornea presented before. Inertia effects are neglected in this problem, due to the typical slow velocity in the application of the loads in this kind of organs. The cornea was discretized with trilinear three-dimensional finite elements. The mesh consisted of 8514 nodes and 7182 elements. A view of the geometry of the model is shown in Fig. 1.

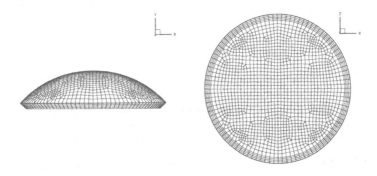

Fig. 1. Geometry of the finite element model for the human cornea

4.1   Palpation of the Cornea

The first test for the proposed technique consists of simulating the palpation of the cornea with a surgical instrument. In order to validate the results, a load was applied to the complete FE model in the central region of the model. The obtained result was compared to the one obtained by employing the model reduction techniques presented before, for the load applied at the same location.

Once the complete model is solved, the most important eigenmodes are extracted from the computed displacements field, together with the initial tangent stiffness matrix. The number of eigenmodes employed in this case was only six, which is, in our experience, the minimum number of modes that should be employed in such a simulation. The modes are depicted in Fig. 2. The associated eigenvalues are, from the biggest to the smallest one, $9.02 \cdot 10^4$, 690, 27, 2.63, 0.221 and 0.0028. As can be seen, the relative importance of these modes in the overall solution, measured by the associated eigenvalue, decreases very rapidly. Note that the reduced model employed only six degrees of freedom, while the complete model employed 8514 nodes with three degrees of freedom each, thus making 25542 degrees of freedom. Of course, if more accurate solutions are needed, a higher number of modes can be employed.

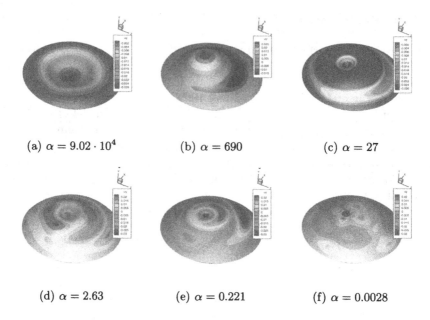

(a) $\alpha = 9.02 \cdot 10^4$ (b) $\alpha = 690$ (c) $\alpha = 27$

(d) $\alpha = 2.63$ (e) $\alpha = 0.221$ (f) $\alpha = 0.0028$

Fig. 2. Six first eigenmodes of the problem employed as global basis for the reduced model simulation

The displacement field obtained for the complete model is compared to that of the reduced model. We chose different positions of the load and compared the results. For a first location of the load, the obtained vertical displacement is shown in Fig. 3.

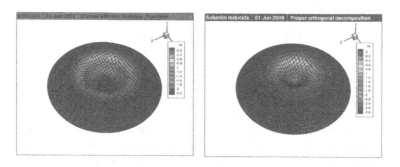

Fig. 3. Vertical displacement field for a first position of the load. Complete model (left) vs. reduced model (right).

The L_2 error norm ranged from very low values (0.08) in the early steps of the simulation, to higher values (around 0.34) for the last step. In our experience, this is a typical upper bound of the obtained error, even if very large deformations are imposed to the simulated organ, as is the case.

The simulations ran at 472-483Hz, which is among the limits imposed by haptic feedback realism, as mentioned before. Of course, the use of more sophisticated, maybe parallelized, codes, could give even faster results.

4.2   Force Prediction

The architecture of a real-time simulator requires, however, the prediction of the response force to a given displacement imposed to the model by means of the haptic device. Thus, a vertical displacement was imposed to node 4144, located more or less in the center of the cornea, with linearly increasing value. While the complete model took around 3 hours to solve the problem, due to the large displacement imposed at the last steps of the simulation, the reduced model still runs at between 400-500 Hz. The results are summarized in Table 1.

Table 1. Error in the predicted force response on the tool. Reduced Order Modelling (ROM) vs. Finite Element Modelling (FEM)

u	$F_{ROM}(N)$	$F_{FEM}(N)$	$Error(\%)$
0.1	0.0045	0.0055	18
0.2	0.0091	0.0108	16
0.3	0.0136	0.0158	14
0.45	0.0204	0.0227	10
0.675	0.0307	0.0321	4
0.9	0.041	0.0405	-1
1.125	0.0511	0.0482	-6
1.35	0.0614	0.0555	-10
1.575	0.0716	0.0628	-14
1.8	0.0818	0.0702	-16
2.025	0.092	0.0779	-18
2.135	0.097	0.0818	-19

As can be noticed, the predicted response is very accurate at the middle of the simulation, and gives some error both at the very beginning of the simulation and for very large strains.

5   Conclusions

In this paper a novel strategy is presented for real-time interactive simulation of non-linear anisotropic tissues. The presented technique is based on model reduction techniques and, unlike previous works [2], it allows for the consideration of both geometrical and material non-linearities.

The reduced models are constructed by employing a set of "high quality" global basis functions (as opposed to general-purpose, locally supported FE shape functions) in a Galerkin framework. These functions are constructed after some direct simulations of the organs performed by standard FE or BE techniques, for instance. These simulations (their tangent stiffness matrices) are made off-line and stored in memory prior to beginning with the real-time simulation.

Results obtained showed good accordance with complete model results, and ran at frequencies of around 400-500 Hz, enough for real-time requirements, even for this very rude code prototypes.

In sum, the technique presented constitutes in our modest opinion an alternative to standard FE simulation techniques for real-time applications involving non-linear and anisotropic materials.

References

1. Alastrué, V., Calvo, B., Peña, E., Doblaré, M.: Biomechanical modeling of refractive corneal surgery. Journal of Biomechanical Engineering-Trasactions of the ASME 128, 150–160 (2006)
2. Barbič, J., James, D.L.: Real-time subspace integration for St. Venant-Kirchhoff deformable models. ACM Transactions on Graphics (SIGGRAPH 2005) 24(3), 982–990 (2005)
3. Basdogan, C.: Real-time simulation dynamically deformable finite element models using modal analysis and spectral lanczos decomposition methods. In: Medicine meets virtal reality 2001, pp. 46–52 (2001)
4. Bro-Nielsen, M., Cotin, S.: Real-time volumetric deformable models for surgery simulation using finite elements and condensation. Computer Graphics Forum 15(3), 57–66 (1996)
5. Cotin, S., Delingette, H., Ayache, N.: Real-time elastic deformations of soft tissues for surgery simulation. In: Hagen, H. (ed.) IEEE Transactions on Visualization and Computer Graphics, vol. 5(1), pp. 62–73. IEEE Computer Society Press, Los Alamitos (1999)
6. Debunne, G., Desbrun, M., Cani, M.-P., Barr, A.H.: Dynamic real-time deformations using space and time adaptive sampling. In: Proceedings of the ACM SIGGRAPH conference, pp. 31–36 (2001)
7. Delingette, H., Ayache, N.: Soft tissue modeling for surgery simulation. In: Ayache, N., Ciarlet, P. (eds.) Computational Models for the Human Body. Handbook of Numerical Analysis, pp. 453–550. Elsevier, Amsterdam (2004)
8. Delingette, H.: Towards realistic soft tissue modeling in medical simulation. Technical Report INRIA RR-3506
9. Karhunen, K.: Uber lineare methoden in der wahrscheinlichkeitsrechnung. Ann. Acad. Sci. Fennicae, ser. Al. Math. Phys. 37 (1946)
10. Loève, M.M.: Probability theory, 3rd edn., Van Nostrand, Princeton, NJ. The University Series in Higher Mathematics (1963)
11. Meier, U., Lopez, O., Monserrat, C., Juan, M.C., Alcaniz, M.: Real-time deformable models for surgery simulation: a survey. Computer Methods and Programs in Biomedicine 77(3), 183–197 (2005)
12. Ryckelynck, D., Chinesta, F., Cueto, E., Ammar., A.: On the a priori Model Reduction: Overview and recent developments. Archives of Computational Methods in Engineering 12(1), 91–128 (2006)

Constitutive Modelling of Endothelium Denudation for Finite Element Simulation of Angioplasty

Sébastien Delorme, Rouwayda El-Ayoubi, and Patricia Debergue

Industrial Materials Institute, 75 de Mortagne Blvd, Boucherville, QC, J4B 6Y4 Canada
{sebastien.delorme, rouwayda.el_ayoubi}@cnrc-nrc.gc.ca
patricia.debergue@cnrc-nrc.gc.ca

Abstract. This study aims at characterizing and modelling the effect of mechanical factors on endothelial denudation during angioplasty, such as normal force between balloon and artery, stretching of arterial walls, and relative displacement between contacting surfaces. Friction damage was applied to porcine aorta samples with different contact forces, relative displacements, and biplanar stretching conditions. After the tests, endothelium denudation was quantified by isolating and counting the remaining endothelial cells. Using multiple-regression analysis, a constitutive model is proposed for integration in finite element software. This model will help optimize balloon and stent deployment conditions to minimize the amount of damage to the endothelium, and eventually to reduce the occurrence of restenosis.

Keywords: angioplasty, restenosis, endothelium, stretch, friction, pressure.

1 Introduction

Intravascular interventions trigger a cascade of cellular and biochemical events often leading to restenosis, a common complication of balloon angioplasty occurring within 6 months of treatment. In cases where stents are implanted, neointimal hyperplasia is the main contributor to a 15 to 20% rate of in-stent restenosis [1],[2].

Neointimal thickness has been shown clinically to be correlated with the aggressiveness of the angioplasty technique, in terms of balloon size and inflation pressure [3], or in terms of vessel wall stretch [4]. It has been shown histologically to be correlated with deepness of vessel injury [5], and with acute luminal stretching [6].

Despite considerable research efforts deployed at trying to inhibit neointimal hyperplasia through systemic or local drug delivery, little effort has been spent on understanding the vessel injury from a mechanical perspective. In vivo animal studies have concluded that denudation of the endothelium alone or in combination with medial injury triggered neointimal hyperplasia [7],[8], but that medial injury alone does not lead to intimal proliferation [9].

Rogers et al. [10] have observed that after the implantation of balloon-deployed stents in rabbit arteries, the endothelium was denuded not only under the stent struts, but also between the so-called stent "cells" (open region between stent struts) because of the balloon protrusion through the stent cells and against the arterial wall. They have not identified whether this denudation was due to friction or contact with the balloon.

F. Bello, E. Edwards (Eds.): ISBMS 2008, LNCS 5104, pp. 19–27, 2008.

Objective. The present study aims at characterizing and modelling the combined effect on the level of endothelium denudation, of the following mechanical parameters: 1) arterial wall stretch, 2) transmural pressure, and 3) friction of the balloon against the arterial wall during its deployment from the folded configuration.

2 Methods

2.1 Experimental Procedure

Descending aortas were harvested from pigs within 8 hours of death. The aortas were cut open longitudinally and cruciform-shaped samples were cut out from each aorta (Fig. 1a). The circumferential and axial directions of the artery were marked on the samples. A total of 51 samples were prepared. All the samples were kept in 0.9% isotonic solution, until the mechanical tests were performed.

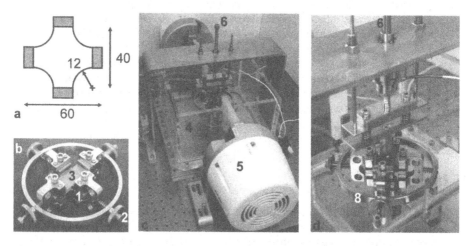

Fig. 1. Experimental setup for friction testing: a) cruciform sample shape and dimensions in mm (gripping areas in grey); b) specimen holder, including grip clamps (1), stretch ratio adjustment screws (2), and knurled plate assembly (3); c) overall view and d) close-up view of friction testing apparatus, including heated bath (4), linear actuator (5), transmural pressure adjustment screw (6) and load cell (7), and specimen holder (8)

Friction tests between polyethylene terephtalate (PET) samples, the material used in many angioplasty balloons, and aorta samples were performed on a custom-designed friction testing apparatus (Fig. 1c, d), using electromagnetic actuators (Bose ElectroForce Systems Group, Eden Prairie, MN). In each test, an aorta sample was mounted with four grip clamps on a specimen holder (Fig. 1b) and a biplanar stretch was applied. To prevent local planar deformation of the sample during the friction test, a cruciform plate assembly was placed over the sample. The plate assembly had a knurled 0.5 mm diamond-shaped pattern on its contact surface to minimize slip between the plate and the artery sample. Plate assemblies of various sizes were designed to exactly fit between the four grip clamps at the selected stretch ratios, to eliminate relative motion between the specimen holder and the assembly.

The specimen holder was then connected to actuators so that endothelial side of the artery was both immerged in saline solution maintained at body temperature and in contact with a PET specimen. The specimen holder had indexed recesses to allow repeatable positioning at predefined orientations. A compressive force was applied over the sample and was maintained constant while the specimen holder was pulled horizontally over a specific distance.

A design of experiment (DOE) was used to investigate the effect of stretch ratio, transmural pressure and friction on the denudation of the endothelium. A 5-factor 3-level Box-Behnken DOE was used to investigate the effect of the following independent variables: applied normal force (F_n; values of 5, 62.5, and 120 N), stretch ratio[1] in the circumferential direction (λ_c; values of 1.00, 1.25, and 1.50), stretch ratio in the axial direction (λ_a; values of 1.00, 1.25, and 1.50), displacement in the circumferential direction (d_c; values of 0.0, 2.25 and 4.5 mm), and displacement in the axial direction (d_a; values of 0.0, 2.25 and 4.5 mm). Vector combinations of axial and circumferential displacements were accomplished by attaching the sample holder to the actuator at different orientations. The contact area was either 15-by-15 mm or 20-by-20 mm depending on the displacement vector. A total of 46 samples were tested, and 5 samples were used as controls.

At the end of each experiment, the endothelial cells (EC) that had detached from the intima were in the saline solution of the heated bath. The aorta sample was taken out of the grips and gently rinsed with saline. A round sample (6 or 10 mm diameter, depending on applied displacements) was punched out of the sample area in contact with the PET and placed on a Petri dish with the luminal side downwards. The intima layer was incubated for 30 min in Phosphate Buffered Saline (PBS) with collagenase (1 mg/ml), in order to detach the remaining EC from the endothelium after the friction tests. The EC were then removed by gently scraping the intimal surface using a plastic cell scraper. The collagenase was eliminated by pelleting the EC at 170 g (1000 rpm) for 5 min. The cell pellet was resuspended in PBS. The EC were stained with antibody anti-Human von Willebrand Factor (vWF). The remaining EC were counted under a microscope with a hematocymeter. The same process was also performed on the control samples. The cell density was obtained by dividing the number of cells counted on a sample by the surface area of the sample. The coverage percentage of EC after a friction test (e_{remain}) was obtained by dividing the cell density on the tested sample by the average cell density on the control samples.

2.2 Shear Deformation Artefact

Preliminary tests have shown that the friction force in the proposed DOE can be high enough to create non-negligible shear deformation of a sample across its thickness, as illustrated in Fig. 2a. To obtain a better estimate of the relative displacement between the two contacting surfaces, the shear deformation was calculated and subtracted from the applied displacement vector (Fig. 2b). The tissue is assumed to be incompressible. Therefore, the following relationship applies to the stretch ratios in the axial, circumferential, and thickness directions (Fig. 2c):

$$\lambda_a \lambda_c \lambda_t = 1 \tag{1}$$

---

[1] The stretch ratio is calculated by dividing the final length of the sample by its initial length.

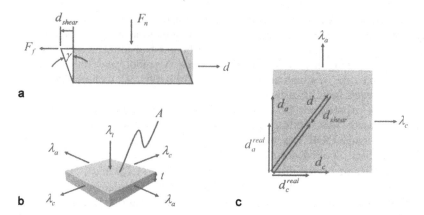

Fig. 2. Shear deformation artefact: a) shear deformation due to friction force; b) three-dimensional stretch on sample; c) vector sum of applied displacement d and shear deformation d_{shear}

Biaxial stretching of the sample reduces its thickness t down to

$$t = \lambda_t t_0 = \frac{1}{\lambda_a \lambda_c} t_0 \,, \tag{2}$$

where t_0 is the initial thickness of the sample. Assuming that the Amonton-Coulomb's law of friction applies, the friction force F_f can be written as a function of the normal force F_n:

$$F_f = \mu F_n \,, \tag{3}$$

where μ is the dynamic friction coefficient. The shear stress τ is given by

$$\tau = \frac{F_f}{A} = \mu \frac{F_n}{A} \,. \tag{4}$$

where A is the contact area. The shear stress can also be expressed in terms of the shear angular deformation γ:

$$\tau = G\gamma \,, \tag{5}$$

where G is the shear modulus. The shear deformation is defined by

$$d_{shear} = \gamma t \,. \tag{6}$$

Substituting equations 2, 4 and 5 into 6 gives

$$d_{shear} = \frac{\mu t_0}{G} \times \frac{F_n}{\lambda_a \lambda_c} \times \frac{1}{A} \,, \tag{7}$$

where $G = 17\text{kPa}$ and $\mu = 0.02$ according to published results [11].

2.3 Statistical Analysis and Modelling

A multiple regression was calculated using NCSS statistical software to model the dependent variable as a function of independent variables. The dependent variable was the percentage of denuded endothelium on the sample, given by

$$e_{denuded} = 1 - e_{remain} \cdot \tag{8}$$

Given that the values of the dependent variable y and of the independent variables x_i can be represented in matrix forms by

$$\mathbf{Y} = \begin{bmatrix} y_1 \\ \vdots \\ y_j \\ \vdots \\ y_N \end{bmatrix}, \quad \mathbf{X} = \begin{bmatrix} 1 & x_{11} & \cdots & x_{p1} \\ \vdots & & & \\ 1 & x_{1j} & \cdots & x_{pj} \\ \vdots & & & \\ 1 & x_{1N} & \cdots & x_{pN} \end{bmatrix}, \tag{9}$$

where N is the number of observations and p is the number of independent variables, as described in [12]. The predicted values of the dependent variables are given by

$$\hat{\mathbf{Y}} = \mathbf{b}'\mathbf{X} . \tag{10}$$

where $\mathbf{b}$ is the coefficient vector calculated by

$$\mathbf{b} = \begin{bmatrix} b_0 \\ b_1 \\ \vdots \\ b_p \end{bmatrix} = (\mathbf{X}'\mathbf{X})^{-1}\mathbf{X}'\mathbf{Y} \cdot \tag{11}$$

3 Results

The following model was calculated using only linear terms involving the five independent parameters (d_a, d_c, F_n, λ_a, λ_c):

$$e_{denuded} = -0.020d_a - 0.046d_c + 5.0 \times 10^{-3} F_n + 0.087\lambda_a + 0.063\lambda_c , \tag{12}$$

where d_a and d_c are expressed in millimetres, F_n is expressed in Newtons, and λ_a and λ_c are dimensionless. The coefficient of determination (R^2) was 0.9523.

For integration in finite element software, a different model is proposed using the following independent variables: the transmural pressure (P), the friction work (W_f), and the sum of biplanar stretch ratios (λ_{1+2}). These independent variables were calculated from the factors of the DOE using the following equations:

$$P = \frac{F_n}{A} , \tag{13}$$

$$W_f = \mu F_n d \, , \tag{14}$$

$$\lambda_{1+2} = \lambda_1 + \lambda_2 = \lambda_a + \lambda_c \, , \tag{15}$$

where d is the length of the displacement vector. Despite the fact that the DOE was not generated with these 3 new independent variables, correlation between the independent variables was low (<0.375). The range of values used in the DOE becomes 0.12 atm (90 mm Hg) to 5.26 atm (4000 mm Hg) for P, 0 to 18.6 mJ for W_f, and 2 to 3 for λ_{1+2} (see Table 1). The following model is proposed to predict endothelium denudation

$$e_{denuded} = 5.5 \times 10^{-4} P + 0.077\lambda_{1+2} + 0.023W_f \, , \tag{16}$$

where P is expressed in kPa and W_f is in mJ. The coefficient of determination was 0.9317. Fig. 3 shows the predicted values using model in equation 13 as a function of the observed values for the dependent variable.

Table 1. Values of the independent variables and the observed dependent variable for each experiment

Exp.	λ_{1+2}	P (kPa)	W_f (mJ)	$e_{denuded}$	Exp.	λ_{1+2}	P (kPa)	W_f (mJ)	$e_{denuded}$
1	2.5	13	0.440	0.000	24	2.5	278	13.243	0.569
2	2.25	156	3.750	0.380	25	2.5	156	6.455	0.496
3	2.25	278	2.292	0.457	26	2.5	156	6.455	0.535
4	2.5	156	9.750	0.535	27	2.5	156	6.455	0.535
5	2.75	156	4.375	0.535	28	3	156	6.913	0.677
6	2.75	156	4.375	0.574	29	2.25	533	2.986	0.462
7	2.5	300	5.270	0.612	30	2.25	533	2.986	0.457
8	2.5	13	0.997	0.070	31	2.75	300	10.666	0.729
9	2.25	278	9.245	0.354	32	2.75	300	10.666	0.690
10	2.25	278	9.245	0.380	33	2.5	13	0.997	0.070
11	2.75	156	11.328	0.569	34	2.25	278	9.245	0.354
12	2.75	156	11.328	0.677	35	2.25	278	9.245	0.354
13	2.5	300	18.620	0.690	36	2.75	156	11.328	0.569
14	2.25	22	0.615	0.070	37	2.75	156	11.328	0.569
15	2.25	22	0.615	0.031	38	2.5	300	18.620	0.651
16	2.75	13	0.628	0.109	39	2.5	13	0.440	0.109
17	2.75	13	0.628	0.031	40	2.25	278	2.292	0.341
18	2	278	3.788	0.302	41	2.25	156	3.750	0.457
19	2.5	278	5.177	0.419	42	2.5	156	9.750	0.612
20	2.5	278	5.177	0.419	43	2.75	156	4.375	0.457
21	2.5	156	6.455	0.457	44	2.75	156	4.375	0.574
22	2.5	156	6.455	0.574	45	2.5	300	5.270	0.612
23	2.5	156	6.455	0.535	46	2.5	156	0.000	0.496

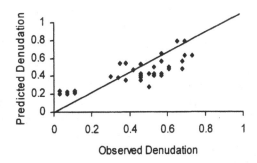

Fig. 3. Predicted vs. observed $e_{denuded}$ using model in equation 13

4 Discussion

In the present study, a constitutive model is proposed for prediction of endothelium denudation. The model is based on in vitro experiments using porcine artery samples. Porcine, rat and rabbit arteries have been used as a model to better understand the mechanical behaviour of arteries and their response to injuries [5],[7],[8],[9], including endothelium denudation, but whether the results of the present study are applicable to humans remains to be demonstrated.

The experiments involved contact between arterial tissue and PET samples. Although care was taken to reproduce in vivo conditions of angioplasty balloon in contact with an artery, it is possible that 1) the surface condition of the balloon might change as the balloon membrane stretches; and that 2) the use of saline solution might create different friction conditions than in an artery filled with blood.

In this study, the speed of relative displacement between the artery and the PET sample was not considered as an independent variable. Slow balloon inflation has been recently shown to reduce neointimal thickening [13],[14]. Whether this effect is due to a reduction of endothelium denudation is unknown. However, recent findings on the effect of strain rate on the mechanical behaviour of the arterial wall [15] rather suggest a relationship between strain rate and overstretch injury.

The experimental method was designed to minimize experimental artefacts. Nevertheless, the biggest artefact appears to be due to the shear deformation of the sample due to the friction force. The magnitude of the artefact was estimated and displacements were corrected accordingly. However, this estimation depends on parameters such as friction coefficient and shear modulus that were obtained from published data rather than being measured on the samples used in the present study. Moreover, the shear modulus was assumed to be constant, although it is possible that the shear modulus varies with the level of stretch of the artery, especially since the stress-strain curve of the artery is nonlinear at the level of stretch used in this study. These simplifying assumptions might impair the accuracy of the model.

Despite the above-mentioned experimental limitation, the selected values for the independent variables generated a wide distribution for the dependent variable, spanning almost the whole range of 0 (intact endothelium) to 1 (no endothelium left), as shown in Fig. 3. This maximizes the range of application of the model.

During an angioplasty and stent implantation, balloon inflation is a combination of two possibly overlapping phases: 1) deployment of the balloon flaps, and 2) stretching of the balloon membrane. During the first phase, friction between the balloon and endothelium can occur if the balloon contacts the arterial wall before the deployment of the balloon flaps is completed. During the second phase, friction between the balloon and endothelium can also occur if the wall does not stretch as uniformly as the balloon. Knowledge of the exact sequence of event requires numerical simulation of the intervention using patient-specific artery shape and mechanical properties.

The proposed constitutive model is expressed as a function of variables that can be calculated in finite element software. Since endothelium denudation is irreversible, at least during balloon deployment, the model must be formulated to take into account the history of the tissue. This is obvious for the friction work, which can never decrease over time. However, it is debatable whether the model should only consider the instantaneous values for the stretch and the pressure or should consider the history of stretch and pressure. It was observed from pilot experiments that pressure alone, without displacement or stretch, causes denudation. Also, when a regression analysis was performed including interactions, first order stretch was a significant factor. So it is proposed that at each load step or iteration of the finite element calculation, the instantaneous values of stretch and pressure and the cumulative values of friction work be used in the calculation of the endothelium denudation. However, in order for the denudation never to decrease, the maximum historical value of the endothelium denudation would always override the instantaneous value, as in

$$e_{denuded}(t) = \max\left[5.5 \times 10^{-4} P(t) + 0.077 \lambda_{1+2}(t) + 0.023 W_f(t)\right]_0^t , \qquad (17)$$

where $P(t)$ and $\lambda_{1+2}(t)$ are instantaneous values, and $W_f(t)$ is a cumulative value given by

$$W_f(t) = \mu \int_{\tau=0}^{t} F_n(\tau) d(\tau) d\tau \cdot \qquad (18)$$

Solid mechanics simulation software with multi-body contact algorithms is currently being developed and used to study the mechanics of balloon-stent-artery interaction [16]. The proposed constitutive model of endothelium damage will be implemented in the software to take into account the complexity of balloon-stent-artery interaction in the patient-specific prediction of endothelium denudation.

5 Conclusion

In this study, a model is proposed to predict endothelium denudation as a function of mechanical variables describing the interaction between balloon and arterial wall. This is the first step towards prediction of arterial damage and biological response of the artery. Other experimental work is underway to develop a model predicting neointimal thickness as a function of stretch, intramural pressure and friction work. Used in finite element analysis of balloon angioplasty and stenting, such models will help clinicians select for a specific patient the balloon size and inflation pressure, not only that maximizes the lumen patency, but also that minimizes arterial injury and long-term lumen loss.

Acknowledgements. This work was supported by the National Research Council of Canada. The authors have no financial interest in the research. We would like to thank Marc-André Rainville for technical help in performing the friction tests.

References

1. Kiemeneij, F., Serruys, P.W., Macaya, C., et al.: Continued benefit of coronary stenting versus balloon angioplasty: five-year clinical follow-up of Benestent-I trial. J. Am. Coll. Cardiol. 37, 1598–1603 (2001)
2. Serruys, P.W., Luijten, H.E., Beatt, K.J., et al.: Incidence of restenosis after successful coronary angioplasty: a time-related phenomenon. A quantitative angiographic study in 342 consecutive patients at 1, 2, 3, and 4 months. Circulation 77, 361–371 (1988)
3. Hoffmann, R., Mintz, G.S., Mehran, R., et al.: Tissue proliferation within and surrounding Palmaz-Schatz stents is dependent on the aggressiveness of stent implantation technique. Am. J. Cardiol. 83, 1170–1174 (1999)
4. Koyama, J., Owa, M., Sakurai, S., et al.: Relation between vascular morphologic changes during stent implantation and the magnitude of in-stent neointimal hyperplasia. Am. J. Cardiol. 86, 753–758 (2000)
5. Schwartz, R.S., Huber, K.C., Murphy, J.G., et al.: Restenosis and the proportional neointimal response to coronary artery injury: results in a porcine model. J. Am. Coll. Cardiol. 19, 267–274 (1992)
6. Kuntz, R.E., Safian, R.D., Carrozza, J.P., et al.: The importance of acute luminal diameter in determining restenosis after coronary atherectomy or stenting. Circ. 86, 1827–1835 (1992)
7. Capron, L., Bruneval, P.: Influence of applied stress on mitotic response of arteries to injury with a balloon catheter: quantitative study in rat thoracic aorta. Cardiovasc. Res. 23, 941–948 (1989)
8. Olson, N.E., Chao, S., Lindner, V., Reidy, M.A.: Intimal smooth muscle cell proliferation after balloon catheter injury. The role of basic fibroblast growth factor. Am. J. Pathol. 140, 1017–1023 (1992)
9. Jamal, A., Bendeck, M., Langille, B.L.: Structural changes and recovery of function after arterial injury. Arteriscler. Thromb. Vasc. Biol. 12, 307–317 (1992)
10. Rogers, C., Tseng, D.Y., Squire, J.C., Edelman, E.R.: Balloon-artery interactions during stent placement: a finite element analysis approach to pressure, compliance, and stent design as contributors to vascular injury. Circ. Res. 84, 378–383 (1999)
11. Caldwell, R.A., Woodell, J.E., Ho, S.P., et al.: In vitro evaluation of phosphonylated low-density polyethylene for vascular applications. J. Biomed. Mater. Res. 62, 514–524 (2002)
12. Multiple Regression. In: NCSS User's Guide, pp. 305-1–305-98 (2007)
13. Umeda, H., Iwase, M., Kanda, H.: Promising efficacy of primary gradual and prolonged balloon angioplasty in small coronary arteries: A randomized comparison with cutting balloon angioplasty and conventional balloon bngioplasty. Am. Heart J. 147, 1–8 (2004)
14. Weiss, T., Leibovitz, D., Katz, I., Danenberg, C., Varshitsky, B., Lotan, C.: The value of computerized angioplasty in patients undergoing coronary stenting: a prospective, randomized trial. Am. J. Cardiol. 100, 118 (2007)
15. Virues Delgadillo, J.O., Delorme, S., DiRaddo, R., Hatzikiriakos, S.: Stiffness of porcine aortas decreases with strain-rate. Simulation in Healthcare 2, 156 (2007)
16. Laroche, D., Delorme, S., Anderson, T., DiRaddo, R.: Computer prediction of friction in balloon angioplasty and stent implantation. In: Harders, M., Székely, G. (eds.) ISBMS 2006. LNCS, vol. 4072, pp. 1–8. Springer, Heidelberg (2006)

Efficient Nonlinear FEM for Soft Tissue Modelling and Its GPU Implementation within the Open Source Framework SOFA

Olivier Comas[1,3], Zeike A. Taylor[2], Jérémie Allard[3], Sébastien Ourselin[2], Stéphane Cotin[3], and Josh Passenger[1]

[1] BioMedIA Lab, The Australian e-Health Research Centre, Brisbane, Australia
[2] University College London, London, United Kingdom
[3] INRIA Alcove, Lille, France

Abstract. Accurate biomechanical modelling of soft tissue is a key aspect for achieving realistic surgical simulations. However, because medical simulation is a multi-disciplinary area, researchers do not always have sufficient resources to develop an efficient and physically rigorous model for organ deformation. We address this issue by implementing a CUDA-based nonlinear finite element model into the SOFA open source framework. The proposed model is an anisotropic visco-hyperelastic constitutive formulation implemented on a graphical processor unit (GPU). After presenting results on the model's performance we illustrate the benefits of its integration within the SOFA framework on a simulation of cataract surgery.

1 Introduction

The field of medical simulation is expanding rapidly. Its multi-disciplinary aspect requires the integration within a single environment of solutions in areas as diverse as visualisation, biomechanical modelling, haptic feedback and contact modelling. This diversity of problems creates challenges for researchers to advance specific areas, and leads rather often to duplication of effort. The Open Source SOFA framework [1] was created to overcome this issue by providing researchers with an advanced software architecture that facilitates the development of new algorithms and simulators. One of the main criteria used for assessing the level of realism within a simulator is the ability for the latter to reproduce the deformation of anatomical structures with high fidelity. Therefore, accurate and efficient soft tissue simulation is a critical concern in surgical simulation. For simplicity, most analyses were initially based on linear formulations in order to satisfy real-time constraints[2,3]. More recently, co-rotational models [4] allowing geometric non-linearities have been introduced to overcome some of the limitations of previous formulations. However, the constitutive law in all these models remain linear. Recently, a nonlinear finite element algorithm was introduced in [5]: the Total Lagrangian Explicit Dynamic algorithm (TLED), which was later implemented on Graphics Processing Units (GPU) [6,7]. This implementation allowed real-time

F. Bello, E. Edwards (Eds.): ISBMS 2008, LNCS 5104, pp. 28–39, 2008.

simulation of soft-tissue deformation with large meshes and a gain in computational performance of more than 16 times what can be achieved on the CPU.

In this paper we describe our CUDA-based re-implementation of the TLED algorithm within the international and Open Source framework SOFA. We show that this integration has a very limited cost in terms of performance by comparing the SOFA version with a standalone implementation. By providing an efficient and accurate nonlinear FEM for soft tissue modelling to worldwide researchers, we thus hope to assist in enhancing the realism of medical simulators. Reciprocally this integration into the SOFA framework benefits from additional features, which we illustrate through an example: the rapid prototyping of a cataract surgery simulator using the TLED algorithm to simulate the deformation of the lens.

2   SOFA, an Open Source Simulation Framework

2.1   Objectives

As previously mentioned, research and development in medical simulation requires many diverse skills. Although their interaction is essential to design a realistic simulator, only few teams have the sufficient resources to build such frameworks. SOFA is an open source framework which aims to answer these challenges. SOFA has been mostly developed by INRIA (the French national institute for research in computer science and control) and CIMIT (Center for Integration of Medicine and Innovative Technology) and is primarily targeted at real-time simulation with an emphasis on medical simulation. SOFA is highly modular and flexible: it allows independently developed algorithms to interact together within a common simulationwhile minimising the development time required for integration [1]. The overall goal is to develop a flexible framework while minimising the impact of this flexibility on the computation overhead. To achieve these objectives, SOFA proposes a new architecture that implements a series of concepts described below.

2.2   SOFA Architecture

High-level modularity. The SOFA architecture relies on the innovative notion of multi-model representation where an object is explicitly decomposed into various representations: Behaviour Model, Collision Model, Collision Response Model, Visual Model and Haptic Model. Each representation can then be optimised for a particular task (biomechanics, collision detection, visualisation, haptics) while at the same time improving interoperability by creating a clear separation between the functional aspects of the simulation components. These representations are then connected together via a mechanism called *mapping.* Various mapping functions can be defined, and each mapping associates a set of primitives of a representation to a set of primitives in the other representation (Figure 1). For instance, a mapping can connect degrees of freedom in a Behaviour Model to vertices in a Visual Model.

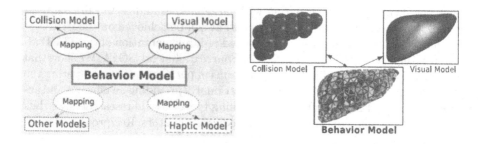

Fig. 1. Multi-model representation in SOFA. Left: a Behaviour Model controls the other representations via a series of mappings. Right: examples of representations with a liver model.

Finer level modularity. In order to easily compare algorithms within SOFA, more flexibility was added to the Behaviour Model by introducing an even finer level of granularity. A series of generic primitives common to most physics-based simulations have been defined: DoF, Mass, Force Field and Solver. The DoF component describes the degrees of freedom, and their derivatives, of the object. The Mass component represents its mass. The Force Field describes both internal and external forces that can be applied to this object. The Solver component handles the time step integration, i.e. advancing the state of the system from time t to time $t + \Delta t$.

Scene-graph. Finally, another key aspect of SOFA is the use of a scene-graph to organise and process the elements of a simulation. Each component is attached to a node of a tree structure. This simple structure makes it easy to visit all or a subset of the components in a scene, and dependencies between components are handled by retrieving sibling components attached to the same node. During the simulation loop, most computations can be expressed as a traversal of the scene-graph. For instance, at each time step, the simulation state is updated by processing all Solver components, which will then forward requests to the appropriate components by recursively sending actions within its sub-tree.

These different functionalities and levels of abstraction allow the user to switch from one component to another by simply editing an XML file, without having to recompile. In particular this permits testing of different computational models of soft tissue deformation, and to assess the pros and cons of various algorithms within the same context.

3   TLED and Its GPU Implementation into SOFA

3.1   TLED Overview

The TLED algorithm is a geometrically and materially nonlinear dynamic finite element method. A nonlinear kinematic framework valid for large deformation is used along with nonlinear constitutive formulations. Both damping and inertial terms are included in the equations of equilibrium, and explicit time integration is employed. A full description of the TLED algorithm can be found in previous publications [5,6,7]. Briefly, the procedure is divided into precomputation and time loop phases. During the simulation loop we proceed as follows:

1. apply imposed displacements and boundary conditions,
2. for each element compute
   (a) displacement derivatives and deformation gradient,
   (b) right Cauchy-Green deformation tensor and 2nd Piola-Kirchhoff stress,
   (c) strain-displacement matrix,
   (d) element nodal force contributions and add these forces to the global nodal forces, and
3. for each node compute new displacements using the central difference method.

3.2   Anisotropic Viscoelastic Constitutive Equation

Viscoelasticity (time dependence) and anisotropy (direction dependence) are well known features of the response of most soft tissues but they are often neglected for computational efficiency. However, because the element stresses are computed directly from strains in explicit analyses, elaborate constitutive models may be incorporated with relative ease. In the present work we used a transversely isotropic visco-hyperelastic model with preferred material direction defined by the unit vector **a**. To the best of our knowledge, this is the first time that such a formulation has been implemented on GPU.

The model is defined in terms of a time-dependent strain energy function

$$\hat{\Psi} = \int_0^t [1 - \alpha(1 - e^{(s-t)/\tau})]\partial_s \Psi^{iso} + \Psi^{vol}, \tag{1}$$

where α and τ are viscoelastic parameters, t is time, and ∂_s denotes partial differentiation with respect to the dummy variable s. Ψ^{iso} and Ψ^{vol} are the isochoric and volumetric components, respectively, of the underlying hyper-elastic strain energy function:

$$\Psi^{iso} = \frac{\mu}{2}(\bar{I}_1 - 3) + \frac{\eta}{2}(\bar{I}_4 - 1)^2, \quad \Psi^{vol} = \frac{\kappa}{2}(J - 1)^2, \tag{2}$$

where μ, η and κ are material parameters J is the determinant of the deformation gradient **F**, $\bar{I}_1$ is the first invariant of the modified right Cauchy-Green tensor $\bar{C} = J^{-2/3}\mathbf{F}^T\mathbf{F}$, and $\bar{I}_4 = \mathbf{a} \cdot \bar{C}\mathbf{a}$ is a pseudo-invariant of $\bar{C}$ and **a**. We consider only viscoelastic isochoric terms.

The 2nd Piola-Kirchhoff stress **S** is obtained through differentiation of (1) with respect to strain. Defining $\Upsilon = \int_0^t 1 - \alpha(1 - e^{(s-t)/\tau})2\partial_{\mathbf{C}_s}\Psi^{iso}ds$ we may obtain $S = 2\partial_{\mathbf{C}}\hat{\Psi} = 2\partial_{\mathbf{C}}\Psi^{iso} + 2\partial_{\mathbf{C}}\Psi^{vol} - \Upsilon$. We may then update stresses at each step n using

$$\mathbf{S}_n = 2(\partial_{\mathbf{C}}\Psi^{iso})_n + 2(\partial_{\mathbf{C}}\Psi^{vol})_n - \Upsilon_n \tag{3}$$

where $(\partial_{\mathbf{C}}\Psi^{iso})_n$ and $(\partial_{\mathbf{C}}\Psi^{vol})_n$ may be computed from the known current deformation $\mathbf{C}_n$. Finally it may be shown that Υ_n is computable from

$$\Upsilon_n = \frac{2\Delta t\alpha(\partial_{\mathbf{C}}\Psi^{iso})_n + \tau\Upsilon_{n-1}}{(\Delta t + \tau)}. \tag{4}$$

3.3   CUDA Description

GPUs achieve a high floating point capacity by distributing computation across a high number of parallel execution threads. They perform optimally as Single Instruction, Multiple Data devices. CUDA is a relatively new C API for compatible NVIDIA GPUs. CUDA organises threads in two hierarchical levels: blocks, which are groups of threads executed on one of the GPU's multiprocessors, and grids, which are groups of blocks launched concurrently on the device, and which all execute the same kernel. Figure 2 represents this thread organisation. As an example, the NVIDIA 8800 GTX used for the results presented in section 4 has 16 multiprocessors, each containing eight processors.

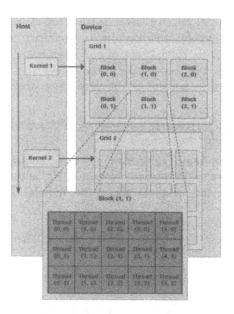

Fig. 2. Each kernel is executed by CUDA as a group of threads within a grid. Image courtesy of NVIDIA [8].

CUDA allows developers to specify the number of threads per block in each execution (the so-called execution configuration), effectively defining the distribution of computational load across all processors. For a given kernel the block dimensions are chosen to optimise the utilisation of the available computational resources. Care should be taken at the multiprocessor level in balancing the available memory required by the kernels with the ability to hide global memory latency. Since a finite amount of memory is available on a multiprocessor, the memory requirements of a kernel will determine how many threads can run concurrently on each. Importantly, CUDA's use of time slicing allows more than one block to be executed concurrently on a single multiprocessor, which has important implications for hiding memory latency. If more than one block is executing, the multiprocessor is able to switch processing between blocks while others are stalled on memory accesses, whereas it has no option but to wait for these if only one block is executing. Therefore for memory bandwidth bound kernels it may be preferable to launch several smaller blocks on each multiprocessor rather than a single larger one if both configurations make the same use of multiprocessor memory resources. While tools are available from NVIDIA for estimating the optimal execution configuration, it has proved necessary to fine tune the configuration experimentally for each kernel.

3.4   GPU Implementation into SOFA

SOFA integration. Implementing a biomechanical model in SOFA translates essentially into writing a new Force Field, i.e. describing the algorithm used to compute internal forces in the model. It merely comes down to creating a single C++ class and changing the position reads and force writes to integrate the algorithm into SOFA's design. The precomputation phase takes place in the initialisation method where relevant variables are computed and passed to an external C function that allocates memory on the GPU and binds textures to it. During the simulation loop, the Solver requests the computation of the forces by launching the appropriate kernels on the GPU.

Kernel organisation. The TLED GPU implementation relies on 2 kernels. The first kernel operates over elements in the model and computes the element stresses based on the current model configuration. It then converts these into nodal force contributions, which are written to global memory. The second kernel operates over nodes and reads the previously calculated element force contributions and sums them for each node. The SOFA central difference solver computes the displacements from the nodal forces. The element force contributions could directly be added to a global force on each node at the end of the first kernel, but this would involve scattered writes. Although CUDA allows scattered writes, it offers no write conflict management between threads, and potential measures to address this severely affect the performance. Therefore, due to the impracticability of scattered writes, the sum operation is reformulated as a gather and the second kernel is needed to sum the nodal forces.

Memory usage. One efficient method for reading global memory data within kernels is texture fetching. Textures may be bound to either *cudaArrays* or regions of linear memory. CudaArrays have been designed to achieve optimal fetching when the access pattern has a high level of 2D locality. In the present application, the access pattern among threads is essentially random (since unstructured meshes are used) and our experiments have shown that texture fetching from linear memory is in fact fastest. Therefore all global memory variables were accessed using this method.

In SOFA, the forces are stored on the GPU in global memory. Since this memory space is not cached, it is important to follow the appropriate access pattern to obtain maximum memory bandwidth, especially given how costly accesses to device memory are. A multiprocessor takes 4 clock cycles to issue one memory instruction for a set of threads. When accessing global memory, there are, in addition, 400 to 600 clock cycles of memory latency. A suboptimal access pattern would yield incoherent writes. The memory bandwidth would then be an order of magnitude lower. In order to prevent this, a key feature of CUDA has been used: *shared memory*. This is a very fast memory shared by all the processors of a multiprocessor. The results of the second kernel are first copied to shared memory and then moved to global memory. If the copies are well organised, it is possible to re-order the access to fulfil all the memory requirements (for both shared and global) and thus reach the maximum bandwidth.

CPU-GPU interaction. CPU-GPU interaction is generally a significant bottleneck in General Purpose GPU applications due to the relatively low interface bandwidth and it is desirable to minimise such interaction. However, interaction cannot be entirely removed from the present implementation since, for example, the solver requires inputs in the form of loaded nodes (which may change due to the interaction with the user) and their displacements, and may need to provide outputs in the form of reaction forces for haptic feedback. CUDA alleviates the problem somewhat by allowing allocation of areas of page-locked host memory which are directly accessible by the GPU and therefore offer much higher bandwidth. In SOFA, all transfers between CPU and GPU are made via this mechanism.

Element technology. Tetrahedral meshes are easily generated and therefore widely used in simulations. Although we used 4-node linear tetrahedra in our previous implementation [6,7], these are known to be susceptible to volumetric locking [9]. To overcome this limitation, we added support for reduced integration 8-node hexahedral elements which are preferable both in terms of accuracy and computational efficiency. A drawback of using hexahedra is the existence of so-called Hourglass modes that have to be addressed to avoid deterioration of the solution [10]. Techniques for suppressing these modes exist, but naturally involve additional computations. However, for a given number of degrees of freedom (DOF), a hexahedral mesh can be built with far fewer elements than a tetrahedral one. Since the majority of the calculations in explicit dynamic analyses are performed per element, this results in reduced overall computation time.

From a GPU perspective, hexahedral element computations are substantially heavier and demand more memory resources. Most of the matrices are twice as large for hexahedra, which necessitates twice as many texture fetches per element, and use of twice as many registers per thread. Thus the occupancy (GPU percentage usage) drops from 25% to only 8%. Similarly, twice as many nodal forces per element are written to global memory by the first kernel. Additional variables for hourglass control are also required. Therefore, on a per element basis hexahedra are significantly less efficient than tetrahedra, especially for GPU execution where memory efficiency is crucial. However, from the point of view of an entire model the lower number of hexahedra required for a given number of degrees of freedom still outweighs this element-wise inefficiency.

4   Results

4.1   Performance of the CUDA-Based Implementation

The algorithm must be efficient to be useful in a real-time environment. Thus we assessed the computational performance of the new CUDA implementation by comparing times to a C++ implementation as we did with our OpenGL-based implementation [6,7]. We generated meshes with between 3 993 and 177 957 DOF and measured the solution times for a single time step on an Intel Core 2 Duo 2.4GHz CPU, 2GB RAM and an NVIDIA GeForce 8800 GTX. Figure 3 shows the substantial speed increase brought by the GPU implementation. We note that the speed improvement factor grows to 53.6×, which makes the CUDA-based implementation approximately 3 times faster than the OpenGL one on current hardware.

4.2   Performance within SOFA

The efficiency and the side-effects of porting the algorithm into the flexible SOFA framework need to be measured. Therefore the performance has been assessed by

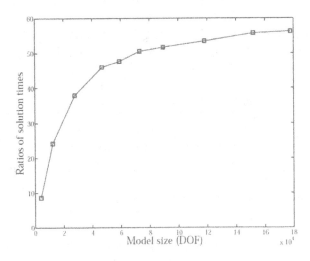

Fig. 3. Ratio of CPU to GPU solution times for the anisotropic viscoelastic formulation

comparing the computational time of the algorithm running within and outside
SOFA. NVIDIA provides a tool to check the GPU implementation by evaluating
many variables during the execution like for instance timings, counts of incon-
sistent reads and writes or GPU occupancy. We used this tool to carry out two
measures:

1. GPU time only estimates the GPU computational time.
2. CPU time allows the evaluation of the execution time with the additional
   overhead due to the framework.

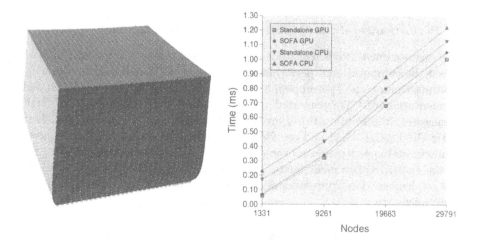

Fig. 4. Left: FEM mesh of cube with 29 791 nodes deformed under a uniformed load.
Right: Comparison of GPU computational timings and CPU overheads between the
SOFA and standalone implementations for different mesh sizes.

The tests were performed on a simple test scene featuring a cube under gravity. Hexahedral meshes with different resolutions from 1 331 to 29 791 nodes were
used and the results are presented in figure 4. In our standalone implementation, the second kernel not only accumulates nodal forces but also adds gravity
and updates positions based on the central difference integration scheme. These
operations are split into separated components in SOFA, in order to introduce
more flexibility (such as applying additional forces or changing the integration
algorithm). While this introduces no noticeable difference on a CPU-based simulation, when using the GPU it is more costly due to overheads in the CUDA
API for the additional kernel launches. Although it reduces the performance by
8.4% for large meshes, this could be optimised away by adding a kernel specific
to a given combination of components.

4.3 A Medical Application: Cataract Surgery

To illustrate the benefit of integrating this new soft tissue deformation model
within the SOFA framework, we created a simulation of phacoemulsification,

used for cataract surgery (figure 5). It consists in removing the natural lens and inserting an artificial intraocular lens implant. To remove the natural lens through a very small opening on the side of the cornea, it is emulsified using an ultrasound device and then aspirated from the eye. During the procedure the lens also undergoes many large deformations which need to simulated. The phacoemulsification step itself is represented by removing elements of the volumetric lens mesh. To be realistic, the simulation requires the lens to be finely meshed, which directly impacts the computation of the deformation of the lens. Limited by the processing resources of the CPU, the previous version of the lens model used a rather coarse mesh to achieve interactive simulation rates. Although studies of biomechanical properties of cornea have shown that the mechanical response was nonlinear [11], viscoelastic [12,13] and anisotropic [11,14], the simulation used a linear model. Therefore the physical model underlying the cataract surgery simulation needed to be improved.

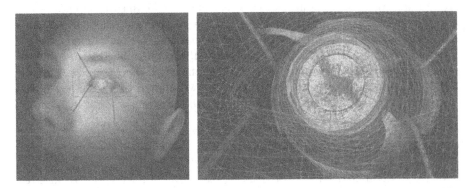

Fig. 5. Cataract surgery simulation. Left: global view of the operating scene. Right: illustration of the complexity of the meshes involved in the simulation.

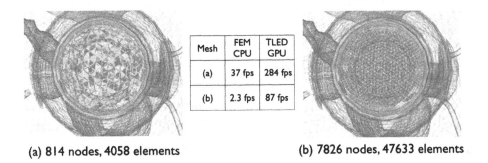

Mesh	FEM CPU	TLED GPU
(a)	37 fps	284 fps
(b)	2.3 fps	87 fps

(a) 814 nodes, 4058 elements   (b) 7826 nodes, 47633 elements

Fig. 6. Comparison of cataract surgery performance using co-rotational FEM on CPU and TLED on GPU

The flexibility of SOFA allows users to easily change the biomechanical models that are used in a simulation by editing the xml file describing the scene. The TLED performance has been compared to a co-rotational [4] FEM implementation on the CPU (see Figure 6). The TLED GPU implementation is clearly faster than the CPU version. However, by using an implicit solver the latter is more robust to collision forces. On the other hand, the TLED adds nonlinearity, viscoelasticity and anisotropy to the mechanical response modelling. This application clearly illustrates the benefits of SOFA. By providing a framework where algorithms can easily be exchanged, one can experiment and ascertain the most suitable combination of algorithms for a particular application.

5   Conclusion

We have implemented an efficient fully nonlinear FEM using CUDA in the open source framework SOFA. Adding a physical solver to SOFA able to model the nonlinear, viscoelastic and anisotropic features of the mechanical response of a material should enhance the fidelity of tissue deformations. We have demonstrated the efficiency of the GPU TLED implementation, and we applied it to cataract surgery simulation. Furthermore we have shown how one can take advantage of SOFA by experimenting with the interaction between algorithms. We hope that our contribution to SOFA will encourage others to share implementations.

References

1. Allard, J., Cotin, S., Faure, F., Bensoussan, P.J., Poyer, F., Duriez, C., Delingette, H., Grisoni, L.: Sofa - an open source framework for medical simulation. In: Medicine Meets Virtual Reality, pp. 13–18 (2007)
2. Cotin, S., Delingette, H., Ayache, N.: Real-time elastic deformations of soft tissues for surgery simulation. IEEE Transactions On Visualization and Computer Graphics 5 (1), 62–73 (1999)
3. Wu, W., Heng, P.: An improved scheme of an interactive finite element model for 3D soft-tissue cutting and deformation. Computer Animation and Virtual Worlds 21 (8-10), 707–716 (2005)
4. Felippa, C.A.: A systematic approach to the element independent corotational dynamics of finite elements. Technical Report CU-CAS-00-03, Center for Aerospace Structures (2000)
5. Miller, K., Joldes, G., Lance, D., Wittek, A.: Total Lagrangian explicit dynamics finite element algorithm for computing soft tissue deformation. Communications in Numerical Methods in Engineering 23(2), 121–134 (2007)
6. Taylor, Z., Cheng, M., Ourselin, S.: Real-time nonlinear finite element analysis for surgical simulation using graphics processing units, pp. 701–708 (2007)
7. Taylor, Z., Cheng, M., Ourselin, S.: High-speed nonlinear finite element analysis for surgical simulation using graphics processing units. IEEE Transactions on Medical Imaging 27(5), 650–663 (2008)
8. NVIDIA Programming Guide 1.1,
   http://developer.nvidia.com/object/cuda.html

9. Hughes, T.J.R.: The Finite Element Method: Linear Static and Dynamic Finite Element Analyses. Prentice-Hall, Inc., Englewood Cliffs (1987)
10. Flanagan, D.P., Belytschko, T.: A uniform strain hexahedron and quadrilateral with orthogonal hourglass control. International Journal for Numerical Methods in Engineering 17, 679–706 (1981)
11. Buzard, K.A., Hoeltzel, D.A.: Biomechanics of the cornea. In: Proceedings of SPIE, vol. 1423, pp. 70–81 (1991)
12. Zeng, Y., Yang, J., Huang, K., Lee, Z., Lee, X.: A comparison of biomechanical properties between human and porcine cornea. Journal of biomechanics (2001)
13. Kobayashi, A., Staberg, L., Schlegel, W.: Viscoelastic properties of human cornea. Journal of experimental mechanics (2006)
14. Elsheikh, A., Brown, M., Alhasso, D., Rama, P., Campanelli, M., Garway-Heath, D.: Experimental assessment of corneal anisotropy. Journal of refractive surgery (2008)

Biquadratic and Quadratic Springs for Modeling St Venant Kirchhoff Materials

Hervé Delingette

ASCLEPIOS Research Project
INRIA Sophia-Antipolis, 2004 route des Lucioles
06902 Sophia-Antipolis, France

Abstract. This paper provides a formal connexion between springs and continuum mechanics in the context of two-dimensional and three dimensional hyperelasticity. First, we establish the equivalence between surface and volumetric St Venant-Kirchhoff materials defined on linear triangles and tetrahedra with tensile, bending and volumetric biquadratics springs. Those springs depend on the variation of square edge length while traditional or quadratic springs depend on the change in edge length. However, we establish that for small deformations, biquadratic springs can be approximated with quadratic springs with different stiffnesses. This work leads to an efficient implementation of St Venant-Kirchhoff materials that can cope with compressible strains. It also provides expressions to compute spring stiffnesses on triangular and tetrahedral meshes.

1 Introduction

Surgery simulation requires the real-time modeling of soft tissue deformation. To obtain a viable compromise between realism and computational speed a wide variety of so-called "physically based models" [7] have been developed. For instance mass-spring models are essentially discrete models that define elastic forces between two vertices based on the variation of the edge length. They are still widely used in surgery simulators because they are supposedly both simple to understand and to implement. Those models however cannot be used to discretize properly two or three dimensional elastic materials based on continuum mechanics. Van Gelder [11] showed for instance that spring-mass models cannot represent linear elastic membranes (see section 2.4 for more discussion). The discrete nature of spring-mass systems is particularly problematic when handling unstructured triangular or tetrahedral meshes. To cope with this limitation, a number of researchers [2] have proposed computational methods to estimate the topologies and the stiffness of springs. A common way to cope with this limitation is to define springs on a rectangular [10,12] lattice in order to improve the isotropy and homogeneity of the deformation.

Several authors have proposed more complex methods than mass-spring models that are to some degree related to continuum mechanics. For instance tensor-mass models [3] correspond to an efficient implementation of linear elastic models while corotational methods [8] extend linear elasticity to large displacements. If

F. Bello, E. Edwards (Eds.): ISBMS 2008, LNCS 5104, pp. 40–48, 2008.

those methods are physically plausible they are not hyperelastic materials and are often limited to small deformations. This is why some authors have used also St Venant Kirchhoff materials for real-time simulation either through optimized data structures [9] or reduced basis [1].

In this paper, we show the *total equivalence* between St Venant Kirchhoff materials defined on discrete surfaces and volumes and a set of *triangular and tetrahedral biquadratic springs*. Those springs do not correspond exactly to the regular springs that are commonly used. However, we show that for small deformations, they are equivalent to regular (quadratic) tensile, angular and volumetric springs, therefore providing analytical expressions of spring stiffness as a function of the rest triangle or tetrahedron geometry. Furthermore, we extend the St Venant Kirchhoff materials to cope with their known limitation in compression. Finally, we show that those biquadratic springs are simple to implement and efficient to compute which makes them well suited for surgery simulation.

2 Membrane Energy on Triangular Meshes

2.1 Membrane Energy

We consider a two-dimensional compact domain $\Omega \subset \mathbb{R}^2$ being deformed into another domain $\Phi(\Omega)$. A physical point $\mathbf{X} \in \Omega$ is moved to a new position $\Phi(\mathbf{X}) \in \Phi(\Omega)$, the function $\Phi(\mathbf{X})$ being the deformation function. The relative change of length around point $\mathbf{X}$ is captured by a two dimensional tensor, the *right Cauchy-Green deformation tensor* $\mathbf{C}$, defined as $\mathbf{C} = \nabla\Phi^T\nabla\Phi$ while the deformation is described by the *Green-Lagrange strain tensor* $\mathbf{E} : \mathbf{E} = 1/2(\mathbf{C}-\mathbf{I})$.

With isotropic St Venant Kirchhoff membrane, there exists a linear relationship between the second Piola Kirchhoff stress tensor and the Green-Lagrange strain tensor. The density of membrane energy $W(\mathbf{X})$ can then be written as : $W(\mathbf{X}) = \frac{\lambda}{2}(tr\mathbf{E})^2 + \frac{\mu}{2}tr\mathbf{E}^2$ where λ and μ are the Lamé coefficients of the material. Those coefficients are simply related to the physically meaningful Young modulus E and Poisson coefficient ν : $\lambda = \frac{E\nu}{1-\nu^2}$ and $\mu = \frac{E(1-\nu)}{1-\nu^2}$. The Young modulus quantifies the stiffness of the material while the Poisson coefficient characterizes the material compressibility ($\nu = 1.0$ for an incompressible surface material and $\nu = 0.5$ for an incompressible volumetric material).

2.2 Membrane Energy of a Deformed Triangle

We now consider that the rest surface Ω is discretized as a set of triangles $\{T_i\}, i = \{1,..,p\}$ and a set of vertices $\{\mathbf{P}_i\}, i \in \{1,..,n\}$. The deformation of that surface is solely determined by the knowledge of the deformed positions of those vertices $\{\mathbf{Q}_i\}, i \in \{1,..,n\}$ such that $\mathbf{Q}_i = \Phi_h(\mathbf{P}_i)$. We write as $\mathcal{A}_P$ (resp. $\mathcal{A}_Q$) the area of the rest triangle T_P (resp. triangle T_Q), l_i (resp. L_i) its edge length and α_i (resp. β_i) its 3 angles (see Figure 1 (Left)).

With a linear triangle element, the deformation function $\Phi(\mathbf{X})$ maps a point $\mathbf{X} \in T_P$ such that $\Phi(\mathbf{X})$ has the same barycentric coordinates in triangle T_Q than $\mathbf{X}$ in triangle T_P :

$$\Phi(\mathbf{X}) = \sum_{i=1}^{3} \eta_i(\mathbf{X})\, \mathbf{Q}_i = \sum_{i=1}^{3} \left(\frac{1}{3} + \mathbf{D}_i \cdot (\mathbf{X} - \mathbf{G}) \right) \mathbf{Q}_i \qquad (1)$$

where $\eta_i(\mathbf{X})$ is the barycentric coordinates of $\mathbf{X}$ in T_P, $\mathbf{D}_i$ are the shape vectors of T_P and $\mathbf{G}$ is the centroid of T_P. The shape vectors $\mathbf{D}_i$ are directed along the inner normal (independently of the triangle orientation) and are of length $1/h_i$, h_i being the altitude of $\mathbf{P}_i$ in T_P. They are algebraically defined in 2D as follows :

$$\mathbf{D}_i = \frac{1}{2A_P}(\mathbf{P}_{i+1} - \mathbf{P}_{i+2})^{\perp} \qquad (2)$$

where $\mathbf{X}^{\perp} = (-y, x)^T$ is the orthogonal of vector $\mathbf{X}$.

From this definition of the deformation, one can easily derive the deformation gradient $\nabla\Phi = \left[\frac{\partial \Phi_i}{\partial x_j}\right] = \sum_{i=1}^{3} \mathbf{Q}_i \otimes \mathbf{D}_i$ and the trace of the right Cauchy Green tensor $tr\mathbf{C} = \sum_{i=1}^{3} \sum_{j=1}^{3} (\mathbf{Q}_i \cdot \mathbf{Q}_j)(\mathbf{D}_i \cdot \mathbf{D}_j)$.

Since C is invariant by translation, one can consider that the origin to be the center $\mathbf{O}_Q$ of the circumscribed circle of T_Q having radius R_Q. Therefore if $i \neq j$, $\mathbf{Q}_i \cdot \mathbf{Q}_j = R_Q^2 - \frac{l_k^2}{2}$ and $\|\mathbf{Q}_i\|^2 = R_Q^2$. Using the property that all shape vectors sum to zero, one gets the following simple expression :

$$tr\mathbf{C} = -\sum_{i=1}^{2}\sum_{j>i}(\mathbf{D}_i \cdot \mathbf{D}_j)l_k^2 = \frac{1}{2A_P}\left(l_1^2 \cot\alpha_1 + l_2^2 \cot\alpha_2 + l_3^2 \cot\alpha_3\right)$$

Finally, by using Heron's formula in triangle T_P which writes the triangle area A_P as a function of square edge length $A_P = \frac{1}{4}\left(L_1^2 \cot\alpha_1 + L_2^2 \cot\alpha_2 + L_3^2 \cot\alpha_3\right)$ one can formulate the 2 invariants of the strain tensor $\mathbf{E}$:

$$tr\mathbf{E} = \frac{\sum_{i=1}^{3} \Delta^2 l_i\, \cot\alpha_i}{2A_P} \qquad tr\mathbf{E}^2 = \frac{\sum_{i\neq j} 2\Delta^2 l_i \Delta^2 l_j - \sum_{i=1}^{3}(\Delta^2 l_i)^2}{64 A_P^2}$$

where $\Delta^2 l_i = (l_i^2 - L_i^2)$ is the square edge elongation.

2.3   Membrane Energy as Triangular Biquadratic Springs

Thus, the total membrane energy to deform triangle T_P into T_Q is a function of square edge variation $\Delta^2 l_i$ and of the angles α_i of the rest triangle :

$$W_{TRBS}(T_P) = \sum_{i=1}^{3} \frac{k_i^{T_P}}{4}(l_i^2 - L_i^2)^2 + \sum_{i\neq j} \frac{c_k^{T_P}}{2}(l_i^2 - L_i^2)(l_j^2 - L_j^2)$$

where $k_i^{T_P}$ and $c_k^{T_P}$ are the tensile and angular stiffness of the *biquadratic springs* :

$$k_i^{T_P} = \frac{2\cot^2\alpha_i(\lambda + \mu) + \mu}{16 A_P} \qquad c_k^{T_P} = \frac{2\cot\alpha_i \cot\alpha_j(\lambda + \mu) - \mu}{16 A_P}$$

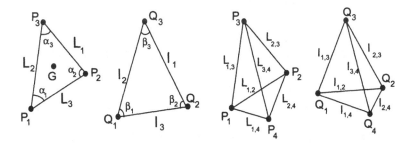

Fig. 1. (Left) Rest triangle T_P having vertices $\mathbf{P}_i$ and its deformed state with vertices $\mathbf{Q}_i$. (Right) Rest tetrahedron T_P and its deformed state T_Q.

We call this formulation the *TRiangular Biquadratic Springs* (TRBS) since the first term can be interpreted as the energy of three *tensile biquadratic springs* that prevent edges from stretching while the second term can be seen as three *angular biquadratic springs* that prevent any change in vertex angles ($c_k^{T_P}$ controling the change in angles around $\mathbf{Q}_i$ and $\mathbf{Q}_j$). Tensile biquadratics springs have a stretching energy which is proportional to the square of the square edge length variation while regular or quadratic springs have an energy related to the edge length variation. The former types of springs are stiffer in extension and looser in compression and are related to the Green-Lagrange strain while the latter are related to the engineering strain [4].

When considering regular triangles, $\alpha_i = \pi/3$, the tensile stiffness is minimal and the angular stiffness is zero for $\nu = 1/3$ and both stiffnesses increase to infinity as ν tends towards 1 (material incompressibility).

2.4 Small Deformations : Triangular Quadratic Springs

When deformations are small, the square edge elongation $\Delta^2 l_i = (l_i^2 - L_i^2)$ can be simplified as $\Delta^2 l_i \approx 2L_i dl_i$ where $dl_i = l_i - L_i$ is the edge elongation. We define the *TRiangular Quadratic Springs* (TRQS) as an approximation of TRBS :

$$W_{TRQS}(T_P) = \sum_{i=1}^{3} \frac{1}{2} \kappa_i^{T_P} (dl_i)^2 + \sum_{i \neq j} \gamma_k^{T_P} dl_i \, dl_j$$

where $\kappa_i^{T_P} = 4L_i^2 k_i^{T_P}$ and $\gamma_k^{T_P} = 4L_i L_j c_k^{T_P}$ are the corresponding tensile and angular spring stiffnesses.

To the best of our knowledge, it is the first time that a theoretical link between spring-mass models and continuum mechanics is established. Without the addition of angular springs, mass springs models with the right stiffness parameters can at best approximate the behavior of a membrane with $\nu = 1/3$ (see [6] for a similar conclusion based on linear elasticity analogy). With angular springs, the TRQS model is equivalent to the TRBS model but only for small deformations of triangles : $dl_i/L_i < 10\%$.

3 St Venant Kirchoff Elasticity on Tetrahedral Meshes

The elastic energy of a St Venant Kirchkoff material is simply expressed as a function of invariant of the Green-Lagrange strain $\mathbf{E}$: $W^{SVK}(\mathbf{X}) = \frac{\lambda}{2}(tr\mathbf{E})^2 + \mu tr\mathbf{E}^2$ where λ and μ are the (volumetric) Lamé coefficients related to Young Modulus and Poisson ratio as follows : $\lambda = \frac{E\nu}{(1+\nu)(1-2\nu)}$ and $\mu = \frac{E}{2(1+\nu)}$.

When discretizing a volumetric body with linear tetrahedral elements, the deformation function $\mathbf{\Phi}(\mathbf{X})$ of a tetrahedron $\mathcal{T}_P$ into tetrahedron $\mathcal{T}_Q$ (see Figure 1 (right)) may be written as a function of the barycentric coordinates $\eta_i(\mathbf{X})$ of $\mathbf{X}$ in $\mathcal{T}_P$ similarly to the 2D case. The 4 shape vectors $\mathbf{D}_i$ are then defined as follows :

$$\mathbf{D}_i = (-1)^i \frac{\mathbf{P}_{i+1} \wedge \mathbf{P}_{i+2} + \mathbf{P}_{i+2} \wedge \mathbf{P}_{i+3} + \mathbf{P}_{i+3} \wedge \mathbf{P}_{i+1}}{6\mathcal{V}_P}$$

Following the same approach as in Section 2.2, we can express the first invariant of the strain tensor as:

$$tr\mathbf{E} = -\frac{1}{2} \sum_{(i,j)\in\mathcal{E}(\mathcal{T})} (\mathbf{D}_i \cdot \mathbf{D}_j)\,(l_{i,j}^2 - L_{i,j}^2)$$

where $\mathcal{E}(\mathcal{T})$ is the set of 6 edges in tetrahedron $\mathcal{T}_P$. The expression of the second invariant $tr\mathbf{E}^2$ is also a function of the variation of square edge length but its complex expression is skipped in this paper due to a lack of space.

From previous results, the total energy to deform tetrahedron $\mathcal{T}_P$ into $\mathcal{T}_Q$ can be derived as a function of square edge variation $\Delta^2 l_i$ and the geometry of the rest tetrahedron :

$$W^{TBS}(\mathcal{T}_P) = \sum_{(i,j)\in\mathcal{E}(\mathcal{T})} \frac{k_{i,j}^{\mathcal{T}_P}}{4}(\Delta^2 l_{i,j})^2 + \sum_{((i,j),(i,k))\in\mathcal{A}(\mathcal{T})} \frac{c_{[i],[j,k]}^{\mathcal{T}_P}}{2} \Delta^2 l_{i,j}\,\Delta^2 l_{i,k} +$$

$$\sum_{((i,j),(k,l))\in\mathcal{O}(\mathcal{T})} \frac{d_{[i,j],[k,l]}^{\mathcal{T}_P}}{2} \Delta^2 l_{i,j}\,\Delta^2 l_{k,l}$$

– $k_{i,j}^{\mathcal{T}_P}$ is a tensile stiffness of a spring linking vertex i and j.

$$k_{i,j}^{\mathcal{T}_P} = \frac{(\lambda+\mu)\mathcal{V}_P}{2}(\mathbf{D}_i \cdot \mathbf{D}_j)^2 + \frac{\mu * \mathcal{V}_P}{2}\|\mathbf{D}_i\|^2\|\mathbf{D}_j\|^2$$

– $c_{[i],[j,k]}^{\mathcal{T}_P}$ is an angular stiffness that prevents the triangle linking vertices i, j and k from changing its angle around vertices j and k. $\mathcal{A}(\mathcal{T})$ is the set of 12 adjacent pair of edges (edge pairs having one and only one vertex in common) in $\mathcal{T}_P$.

$$c_{[i],[j,k]}^{\mathcal{T}_P} = \frac{(\lambda+\mu)\mathcal{V}_P}{2}(\mathbf{D}_i \cdot \mathbf{D}_j)(\mathbf{D}_i \cdot \mathbf{D}_k) + \frac{\mu\mathcal{V}_P}{2}\|\mathbf{D}_i\|^2(\mathbf{D}_j \cdot \mathbf{D}_k)$$

- $d_{[i,j],[k,l]}^{\mathcal{T}_P}$ is a volumetric stiffness that prevents the tetrahedron $\mathcal{T}_P$ from collapsing. $\mathcal{O}(\mathcal{T})$ is the set of 3 opposite pair of edges in $\mathcal{T}_P$.

$$d_{[i,j],[k,l]}^{\mathcal{T}_P} = \frac{\lambda \mathcal{V}_P}{2}(\mathbf{D}_i \cdot \mathbf{D}_j)(\mathbf{D}_k \cdot \mathbf{D}_l) + \frac{\mu \mathcal{V}_P}{2}((\mathbf{D}_i \cdot \mathbf{D}_k)(\mathbf{D}_j \cdot \mathbf{D}_l) + (\mathbf{D}_i \cdot \mathbf{D}_l)(\mathbf{D}_j \cdot \mathbf{D}_k))$$

Thus the elastic energy can be decomposed into the sum of $6 + 12 + 3 = 21$ terms. We call this formulation of the St Venant Kirchoff energy, the *Tetrahedral Biquadratic Springs* (TBS).

Similarly to the membrane energy, each biquadratic springs may be approximated by a quadratic spring by making the following hypothesis : $\Delta^2 l_{i,j} \approx 2L_{i,j}(l_{i,j} - L_{i,j}) = 2L_{i,j}dl_{i,j}$. Typically those approximations are valid for small strains, *i.e.* when $|dl_{i,j}/L_{i,j}| < 0.1$. Under this hypothesis, we can define the *Tetrahedral Quadratic Springs* (TQS) as an approximation of TBS :

$$W^{TQS}(\mathcal{T}_P) = \sum_{(i,j)\in\mathcal{E}(\mathcal{T})} \frac{\chi_{i,j}^{\mathcal{T}_P}}{4}(dl_{i,j})^2 + \sum_{((i,j),(i,k))\in\mathcal{A}(\mathcal{T})} \frac{\gamma_{[i],[j,k]}^{\mathcal{T}_P}}{2} dl_{i,j} \, dl_{i,k} +$$

$$\sum_{((i,j),(k,l))\in\mathcal{O}(\mathcal{T})} \frac{\xi_{[i,j],[k,l]}^{\mathcal{T}_P}}{2} dl_{i,j} \, dl_{k,l}$$

where the TQS tensile, angular and volumetric stiffnesses are proportional to their TBS counterpart :

$$\chi_{i,j}^{\mathcal{T}_P} = 4L_{i,j}^2 \, k_i^{\mathcal{T}_P} \quad \gamma_{[i],[j,k]}^{\mathcal{T}_P} = 4L_{i,j}L_{i,k} \, c_{[i],[j,k]}^{\mathcal{T}_P} \quad \xi_{[i,j],[k,l]}^{\mathcal{T}_P} = 4L_{i,j}L_{k,l} \, c_{[i,j],[k,l]}^{\mathcal{T}_P}$$

4   Compressible St Venant Kirchhoff Materials

The elastic energy of St Venant Kirchhoff material only depends on the invariants of $\mathbf{E}$ which implies that this elastic energy is invariant with respect to any rotations, translations but also any plane reflexion. Therefore, it is easy to show that the elastic energy becomes locally minimum when a triangle or tetrahedron becomes flat. This entails that zero elastic force is applied to resist the collapse of a triangle or tetrahedron. This severely limits the use of St Venant Kirchoff materials since it is not appropriate to simulate materials under compression.

There exists different ways to cope with this limitation. If $J = \det \boldsymbol{\Phi}$ is the ratio of the deformed volume over the rest volume, then one option is to add an energy term proportional to $(J-1)^2$ [9] or to $\log|J|$ as done for Neo Hookean materials. Also some authors have proposed to cope for flat or inverted elements at the cost diagonalizing the deformation gradient matrix with a polar [5] or QR [8] decomposition .

In order to improve the compressibility of St Venant Kirchhoff (SVK) materials while limiting the added computational cost, we propose to add the following energy W^{comp} term to the energy : $W^{comp} = 0$ if $J > 1$ and $W^{comp} = (\lambda + \mu)(J-1)^4/2$ if $J < 1$. Figure 2 (Left) compares the elastic force on a single tetrahedron when a vertex is moved toward its opposite triangle, F^{SVK} being

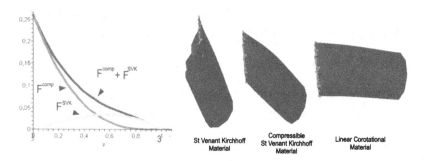

Fig. 2. (Left) Effect of the added volume penalty term F^{comp} on the compression force computed on a single tetrahedron; (Right) Compressible constraint allows to compute the large bending of cylinder under its own weight

the regular SVK force and F^{comp} being the additional force. The composite force $F^{comp} + F^{SVK}$ is C^1 continuous and is a cubic function of strain similarly to the extension case. Compression strain of more than 50% can be reached with this formulation. Figure 2 (Right) shows the effect of the additional compression energy on a soft cylinder clamped on one side and bending upon its own weight. With the regular SVK material, several tetrahedra collapse due to the amount of compression (thus falling into a local minimum) but not for the compressible SVK. Furthermore, the compressible SVK bends significantly but much less than a corotational material having the same Lamé coefficients ($E = 1$, $\nu = 0.35$). This shows that corotational models have a far greater resistance to shear stress than SVK models.

5   Results and Implementation Issues

In this paper, we provide analytical stiffness parameters $\kappa_{i,j}^{T_P}$ and $\chi_{i,j}^{T_P}$ involving Young Modulus and Poisson ratios for spring mass-models defined on triangulation and tetrahedral meshes. We have verified that those spring-mass models behave consistently whether the mesh is regularly tesselated or not [4]. Their behavior is however physically plausible only on triangulations for a choice of the Poisson coefficient equal to 1/3. To obtain reasonable results for other Poisson coefficient, it is required to implement the full TRQS and TQS models that involve angular and volumetric springs.

The TBS, TQS, TRBS and TRQS models have been implemented in the SOFA platform, elastic force and tangent stiffness matrices being analytically derived from the elastic energy presented in this paper. In terms of performance, preliminary studies show that the TBS and TRBS are slightly faster than the TRQS and TQS, since the former do not involve any square root evaluation and lead to simpler expression. Figure 3 (Top) provides some benchmark for the elastic force evaluation and stiffness matrix-vector product on a triangulation computed on a laptop computer with a Core Duo T2400 1.83 MHz processor.

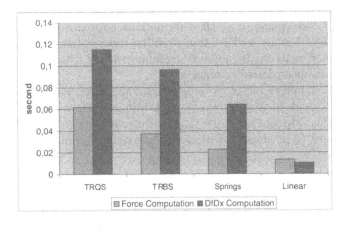

Fig. 3. (Top) Execution times of the elastic force and the matrix-vector product for 4 different elastic membrane formulations; (Bottom) Liver surgery simulation performed in real time with a compressible TBS model before and after resection

In this implementation, edges, triangles and tetrahedra are successively scanned to apply respectively tensile, angular and volumetric forces. It appears that the biquadratic model is only 60% more expensive than the spring-mass model, while the quadratic model is itself 60 % more expensive than the biquadratic model. On tetrahedral meshes, first implementations seem to indicate similar trends, with the TQS model being twice slower than the TBS, and in the worst case (when all tetrahedra are compressed) the compressible TBS being 65% slower than the regular TBS model. Corotational models seem also be 80% slower than the regular TBS. Figure 3 (Bottom) show the use of a compressible TBS model for surgery simulation on a tetrahedral mesh with 1059 vertices and 5015 tetrahedra. Interactive simulation of resection was obtained using an implicit Euler time integration. Edge, triangle and tetrahedra stiffnesses are updated each time tetrahedra are being removed.

6   Conclusion

In this paper, we have first showed the equivalence between surface and volumetric St Venant Kirchhoff materials ad triangular and tetrahedral biquadratic springs. Furthermore we showed that quadratic tensile, angular and volumetric spring models correspond to small deformations approximations of those biquadratic springs. Finally we have improved the compression behavior of the St Venant Kirchhoff material by adding a volume penalty which makes the compression and extension forces symmetric.

Future work should include a more thorough comparison with existing FEM or corotational methods both in terms of computational performance and behavior. Extension of the proposed approach to anisotropic and incompressible materials is also important to produce realistic soft tissue deformation.

References

1. Barbič, J., James, D.L.: Real-time subspace integration for St. Venant-Kirchhoff deformable models. ACM TOG (SIGGRAPH 2005) 24(3), 982–990 (2005)
2. Bianchi, G., Solenthaler, B., Székely, G., Harders, M.: Simultaneous topology and stiffness identification for mass-spring models based on fem reference deformations. In: Barillot, C., Haynor, D.R., Hellier, P. (eds.) MICCAI 2004. LNCS, vol. 3216, pp. 293–301. Springer, Heidelberg (2004)
3. Cotin, S., Delingette, H., Ayache, N.: A hybrid elastic model allowing real-time cutting, deformations and force-feedback for surgery training and simulation. The Visual Computer 16(8), 437–452 (2000)
4. Delingette, H.: Triangular springs for modeling nonlinear membranes. IEEE Transactions on Visualization and Computer Graphics 14(2), 329–341 (2008)
5. Irving, G., Teran, J., Fedkiw, R.: Tetrahedral and hexahedral invertible finite elements. Graph. Models 68(2), 66–89 (2006)
6. Lloyd, B., Szekely, G., Harders, M.: Identification of spring parameters for deformable object simulation. IEEE Transactions on Visualization and Computer Graphics 13(5), 1081–1094 (2007)
7. Nealen, A., Muller, M., Keiser, R., Boxerman, E., Carlson, M.: Physically based deformable models in computer graphics. Technical report, Eurographics State of the Art, Dublin, Ireland (September 2005)
8. Nesme, M., Payan, Y., Faure, F.: Efficient, physically plausible finite elements. In: Dingliana, J., Ganovelli, F. (eds.) Eurographics (short papers) (august 2005)
9. Picinbono, G., Delingette, H., Ayache, N.: Non-Linear Anisotropic Elasticity for Real-Time Surgery Simulation. Graphical Models 65(5), 305–321 (2003)
10. Terzopoulos, D., Platt, J., Barr, A., Fleischer, K.: Elastically deformable models. In: Computer Graphics (SIGGRAPH 1987), vol. 21, pp. 205–214 (1987)
11. Van Gelder, A.: Approximate simulation of elastic membranes by triangulated spring meshes. Journal of Graphics Tools: JGT 3(2), 21–41 (1998)
12. Waters, K., Terzopoulos, D.: Modeling and animating faces using scanned data. The Journal of Visualization and Computer Animation 2, 129–131 (1991)

Exploring Parallel Algorithms for Volumetric Mass-Spring-Damper Models in CUDA

Allan Rasmusson[1,2], Jesper Mosegaard[3], and Thomas Sangild Sørensen[1,4]

[1] Department of Computer Science University of Aarhus, Denmark
[2] Center for Histoinformatics, University of Aarhus, Denmark
[3] Alexandra Institute, Denmark
[4] Institute of Clinical Medicine University of Aarhus
{alras,mosegard,sangild}@daimi.au.dk

Abstract. Since the advent of programmable graphics processors (GPUs) their computational powers have been utilized for general purpose computation. Initially by "exploiting" graphics APIs and recently through dedicated parallel computation frameworks such as the Compute Unified Device Architecture (CUDA) from Nvidia. This paper investigates multiple implementations of volumetric Mass-Spring-Damper systems in CUDA. The obtained performance is compared to previous implementations utilizing the GPU through the OpenGL graphics API. We find that both performance and optimization strategies differ widely between the OpenGL and CUDA implementations. Specifically, the previous recommendation of using implicitly connected particles is replaced by a recommendation that supports unstructured meshes and run-time topological changes with an insignificant performance reduction.

Keywords: Mass-Spring-Damper Models, GPGPU and Deformable Models.

1 Introduction

In the past many biomechanical models have been suggested to simulate soft tissue deformations [1]. Most popular are the finite element models (FEMs) and mass-spring-damper models (MSDMs). Of these models the non-linear FEMs provide the most accurate description of the tissue behavior [2][3]. They require however a significant amount of computation which can be prohibitive in many real-time applications. At the expense of precision, particularly for large deformations, linearized FEMs have been suggested to alleviate this problem [4][5]. MSDMs combine non-linear tissue characteristics *and* fast computation. Due to these properties the MSDMs have been widely used to simulate tissue deformation in existing surgical simulators.

Commodity graphics hardware (GPUs) is an emerging platform for general purpose computation (GPGPU). It is increasingly utilized to accelerate the computation of biomechanical models [6][7][8]. The results obtained in these references were accomplished by exploiting well-known graphics APIs such as

F. Bello, E. Edwards (Eds.): ISBMS 2008, LNCS 5104, pp. 49–58, 2008.

OpenGL or DirectX. This is, unfortunately, a cumbersome process with a steep learning curve. In order to make GPGPU more accessible, one of the major hardware vendors, NVIDIA, recently released a new framework named CUDA - Compute Unified Device Architecture [9].

The purpose of this paper is to investigate which benefits the CUDA framework offers over OpenGL (or equivalently DirectX) in the computation of MSDMs, particularly with respect to speed and functionality. We evaluate multiple parallel implementation strategies for the MSDMs in CUDA and compare their performance to highly optimized OpenGL code [7]. Specifically, we evaluate whether the conclusions derived in [7] can be transferred from OpenGL to CUDA.

2   Theory

2.1   CUDA Overview

The CUDA framework exposes the multiprocessors on the GPU for general purpose computation through a small number of simple extensions of the C programming language. Compute intensive components of a program can be offloaded to the GPU in so-called *kernels*, each of which can be executed in parallel on different input data. This is known as SIMD, Single Instruction Multiple Data. A configuration of *threads* that will execute the kernel in parallel is specified as *blocks* of threads with constant width, height and depth. As the maximum number of threads in a block is limited (currently to 512), multiple blocks are distributed in a rectangular *grid* in order to obtain the desired number of threads. CUDA maps this grid to the GPU such that each multiprocessor executes one or more blocks of threads.

To achieve optimal performance it is important to minimize the cost of memory accesses in a kernel. This is achieved through careful utilization of the different memory pools, which are depicted in Fig. 1. The figure illustrates the small amount of on-chip shared memory which can be used within a block for inter-thread communication. Utilizing this shared memory efficiently yields memory accesses as fast as register accesses (2 clock cycles). In contrast, the main device memory accessible by all threads, has a worst-case access time of 400-600 clock cycles. Finally, Fig.1 shows the texture memory which can be utilized for very fast cached access to any give subset of the device memory.

As any CUDA-device is able to read 32-bit, 64-bit or 128-bit in a single instruction it is furthermore important to organize data in device memory to utilize this. In particular data should be aligned to the appropriate 4, 8 or 16 byte boundaries, even for data of sizes not matching 32-bit, 64-bit or 128-bit. If three floats are needed, performance is optimized by storing four properly aligned floats which can be read in a single 128-bit memory, contrary to storing only three float which are accessed using a 64-bit *and* a 32-bit memory access. The cost is the extra memory needed for padding.

The real benefit is, however, when threads executed in parallel access a contiguous part of the device memory. If consecutive threads access consecutive

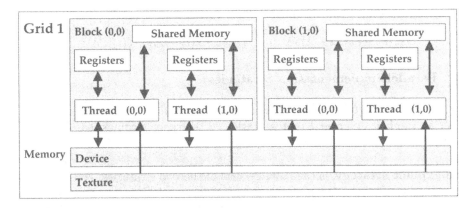

Fig. 1. CUDA Memory Model, adapted from [9]. The arrows indicate read or write access for a thread, ie. textures are *non*-writable. In this example the configuration of threads is a 2×1 grid of 2×1 blocks resulting in 4 threads in total.

memory address, say 32-bit floats, the individual memory instructions are replaced by a *single* memory access. This is known as memory coalescence and is at present possible for groups of 32 consecutive threads.

2.2   Mass-Spring-Damper Model (MSDM)

In a mass-spring system particles p_i, $i = 1 \dots N$ with masses m_i are interconnected by springs. Every particle is displaced by the forces induced by its interconnecting springs. The relation is described by a second order differential equation according to Newton's second law of motion:

$$m_i \mathbf{a}_i = \sum_j \mathbf{f}_{ij}, \tag{1}$$

where $\mathbf{a}_i$ is an acceleration vector for particle i and the force $\mathbf{f}_{ij}$ along a spring between particle i and particle j is expressed using Hookes Law:

$$\mathbf{f}_{ij} = k_{ij}(l_{ij} - \|\mathbf{x}_i - \mathbf{x}_j\|)\frac{\mathbf{x}_i - \mathbf{x}_j}{\|\mathbf{x}_i - \mathbf{x}_j\|}. \tag{2}$$

The term l_{ij} constitutes the rest length of the spring, and k_{ij} is the spring constant that determines the elastic property of the spring. The non-linearity of the MSDM stems from the term $\|\mathbf{x}_i - \mathbf{x}_j\|$.

One method to numerically integrate equation (1) is the *Verlet* integration scheme. It is a particularly good choice as the updated position for each particle is calculated solely from the force vector $\mathbf{f}$ and the particle's two previous positions. Moreover, it can easily be computed for each particle in parallel. The Verlet integration is given by:

$$\mathbf{x}(t + \varDelta t) = 2\mathbf{x}(t) - \mathbf{x}(t - \varDelta t) + \frac{\mathbf{f}(t)}{m}(\varDelta t)^2 \tag{3}$$

Artificial damping can be introduced in an ad-hoc manner [10].

$$\mathbf{x}(t + \Delta t) = (2 - \lambda) \cdot \mathbf{x}(t) - (1 - \lambda) \cdot \mathbf{x}(t - \Delta t) + \mathbf{f}(t)(\Delta t)^2. \qquad (4)$$

2.3   Parallel Implementation Strategies

As presented in [7], on the gpu there are two implementation strategies utilizing either an *implicit* or *explicit* representation of the springs connected to a given particle.

The explicit strategy is the more general of the two strategies. For each particle it maintains a list of indices to particles to which it is connected. Two memory accesses are required to look up the position of each neighboring particle; one memory access to acquire the index of the neighbor particle, and one further memory access to determine the position of the particle at that index. Using this data structure it is straight-forward to represent any given connectivity of the MSDM.

The implicit strategy on the other hand requires datasets in which the particles are located in a regular three-dimensional grid (or structured mesh). Here, a particle can be connected to its neighboring particles only. For any particle, the addresses of neighbor positions can now be calculated by using a fixed set of constant offsets. Hence only one memory access is required to retrieve the position of each neighbor particle. To represent arbitrary morphology, particles are marked either active or inactive and only the active particles are considered part of the desired morphology.

3   Methods

This section describes how the strategies presented in section 2.3 for solving the MSDM have been implemented in CUDA.

The basic layout of test data used in this paper is a grid of particle positions represented as a 3D float array. Only adjacent particles in the 3D grid are connected by springs. Spring rest lengths are thus fixed as illustrated in Fig. 2.

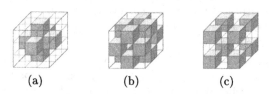

(a)   (b)   (c)

Fig. 2. Categorization of neighbors for a particle in the center of a $3 \times 3 \times 3$ cube. (a) rest length 1, (b) rest length $\sqrt{2}$ and (c) rest length $\sqrt{3}$. The configurations shown denote spring groups 1, 2 and 3 respectively.

3.1 Implicit Addressing

The computation associated to each particle, both active and inactive, is addressed by a single thread in CUDA. From the thread and block ids in the grid a mapping is established to the corresponding memory address of the particle. A neighbor mask (bitpattern) is used to differentiate active and inactive neighbors. Figure 3 shows an example of the bitpattern for a particle which is connected to three neighbors. Each bit is tested iteratively using logical operators. If the entire neighbor mask is zero, the particle corresponding to the current thread is an inactive particle. Hence, a single memory access determines whether a particle is active or inactive.

Fig. 3. Neighbor mask for a particle connected to neighbors 1,2 and 10. Spring group 1 is represented as the lowest 6 bits for which it is known that the rest length is equal to 1. Similarly, bits 6-17 represent spring group 2 with a rest length of $\sqrt{2}$. Spring group 3 is omitted in this example.

The two most recent arrays of the particle positions are required to compute equations (2) and (4). These arrays should be accessed either using cached device memory or shared memory. Memory access can easily be cached, whereas utilizing shared memory requires us to manually handle memory transfer to shared memory from device memory.

Shared Memory Implementation: The "rectangular" subset of device memory corresponding to the threads (particles) in the current block is initially copied from device memory to shared memory from where it is subsequently accessed. The shared memory layout is padded with an additional layer of particles in order for the threads (particles) at the border of the block to access their neighbor particles corresponding to a different block. This is illustrated in figure 4. To access neighbors of different depth, three rectangles like the one depicted in figure 4(b) is copied to shared memory for each block initially in the kernel. This strategy limits access to device memory to a few reads per particle.

Memory Coalescence: The particles are represented in memory as three floats for position and a 32-bit word for the neighbor mask. As it is possible to encode the 32-bit neighbor mask in a float, all information for a particle fits neatly into a float4 datatype. By using a block width to a multiple of 32, the criteria for memory coalescence is supported for all groups of 32 threads executed in parallel.

3.2 Explicit Addressing

The basic layout of data for the explicit addressing strategy is a *compact* array of particle positions containing only active particles. This array is accessed from

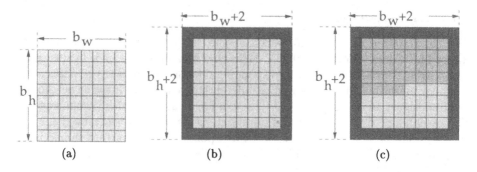

Fig. 4. Copying particle positions from device to shared memory. (a) depicts the block of particles of width b_w and height b_h, (b) shows the additional red frame of neighbor particles, (c) the blue particles are responsible of reading the red frame into shared memory.

	P_{i-1}	P_i	P_{i+1}
	index	index	index
	counter	counter	counter
	n1	n1	n1
	n2	n2	n2
	⋮	⋮	⋮
	nmax	nmax	nmax

Fig. 5. Explicit connections for particles p_{i-1}, p_i and p_{i+1}

device memory using a 1D cache. Shared memory is not used due to the lack of locality in general unstructured meshes.

For each thread a list containing the particle indices of the corresponding particle and all connected particles is stored explicitly as illustrated in Fig. 5. The list has a predetermined maximum number of spring entries to allow easy indexing into an array of lists based on the thread and block id. Furthermore, a counter is included in each list to indicate the number of springs actually present. This data structure makes run-time changes to the topology very easy to support, i.e. to erase or add springs due to cutting or suturing.

A secondary list is maintained to provide per spring parameters such as spring stiffness and rest lengths. In the special case that particles are known to be laid out in a regular grid/structured mesh, this secondary list can be omitted; spring rest lengths can be determined from the corresponding spring group indicated by an individual counter per group. In this special case we furthermore assume that the spring stiffness is constant for all springs.

Memory Coalescence: Memory coalescence is not achieved if the neighbor lists are simply concatenated and stored in memory. The reason is that there is a separation of $32*(nmax+2)$ bits between the individual words read in parallel by the threads. Ie., the necessary contiguous memory layout is not present. In order

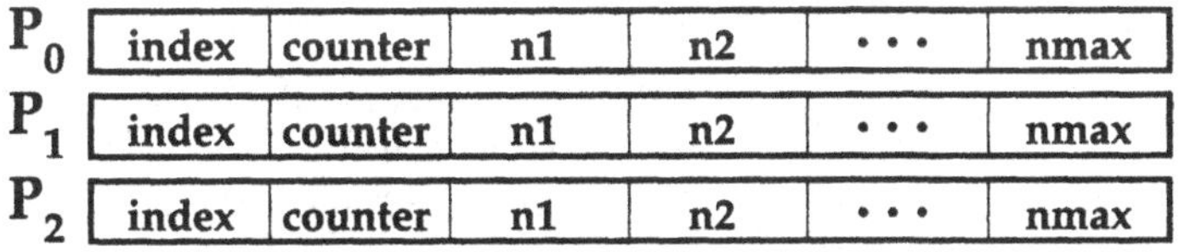

(a) Data layout *not* supporting memory coalescence since indices, counters and neighbor addresses are store linearly for each thread.

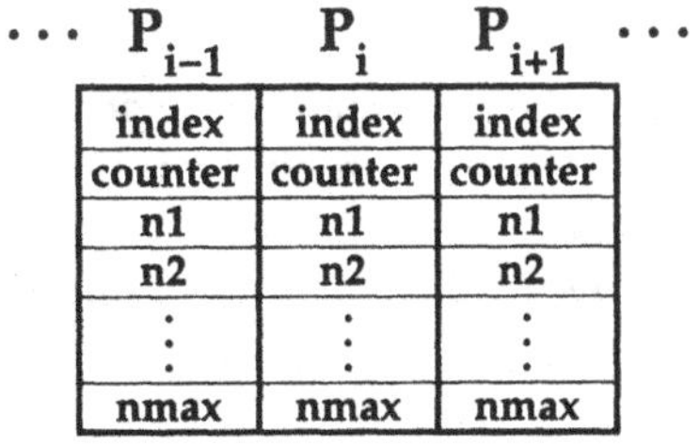

(b) Data layout supporting memory coalescence. By storing the neighbor lists "vertically" yields consecutive indices, counters and neighbor addresses for threads in parallel.

Fig. 6. Memory layout ensuring coalesced access for the explicit strategy

to have consecutive threads read consecutive memory addresses, it is necessary to first store all 32-bit indices, followed by the counter and the neighbor addresses. In a sence, the lists are stored "vertically". This is illustrated in Fig. 6. Again, a block width of a multiple of 32 fullfills the memory coalescence criterias for all groups of 32 threads.

3.3   Test Setup

To test the SMDM implementations we selected two datasets; a compact box and a realistic morphological dataset of a heart obtained from 3D MRI and previously used in a cardiac surgery simulator [11]. Both datasets, listed in Tab. 1, are stored as a binary three-dimensional grid indicating whether a voxel is part of the morphology or not. This data layout can be used by both the implicit and explicit layout and is thus suitable for comparison. The box dataset is optimal for the implicit method since no inactive particles are present.

Consistent with the OpenGL implementation to which we intend to compare performance [7], all implementations use a fixed connectivity pattern consisting of spring groups 1 and 2.

Furthermore, tests were made on boxes of different sizes in order to investigate how performance scales with increasing input size. This was done only for the explicit implementations as the ratio of active vs. inactive particles is irrelevant for this method.

The test platform was Windows XP 32bit, AMD 64 FX55 at 2.61Ghz on MSI K8N-Diamond, PCI-Express x16, GeForce 8800 GTX GPU (768 MB RAM), CUDA Release 1.1, Nvidia Forceware v. 169.21.

Table 1. Technical data about datasets

Dataset	Total Particles	Active Particles	Springs	Ratio Active/Total
Heart	140760	29449	220495	0.209
Box	32768	32768	280240	1.00

4   Results

Results of simulation runs for the described strategies are summarized in Fig. 7. For comparison results of an OpenGL implementation of the implicit strategy are also included.

The execution times for the explicit methods run on boxes of varying sizes are shown in Fig. 8.

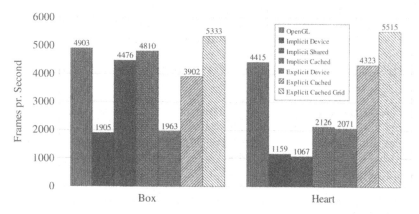

Fig. 7. Chart of performance for the original OpenGL and the implicit and explicit CUDA implementations. For the implicit method three memory strategies (device, shared, and cached) have been tested, while the explicit method is tested for device memory access and cached access. The cached is further divided into a general and a version optimized for grids, the "Explicit Cached Grid".

5   Discussion

In this paper we investigated various GPU implementations of the MSDM using CUDA. The purpose was to evaluate whether previously published results utilizing OpenGL for GPGPU could be reproduced or even improved, and to evaluate if the recommended strategies obtained using OpenGL transfers to CUDA.

Our results (Fig. 7) showed that the fastest CUDA implementation, the strategy utilizing cached explicit memory addressing, outperforms or performs equally well as the fastest OpenGL implementation.

In [7] it was reported that the OpenGL implementation using explicit addressing achieves only 50% of the performance obtained using implicit addressing due to the extra memory indirection. For this reason implicit addressing was recommended over the explicit version. Using CUDA we notice however that this recommendation must be "inverted" to instead recommend explicit addressing. This change in recommendation has several positive side effects, most interestingly added flexibility and ease of implementation. The explicit method is very flexible since it can represent arbitrary geometry and easily allows for run-time changes of topology. Moreover, it turns out that the simplest of the implemented strategies performs superiorly.

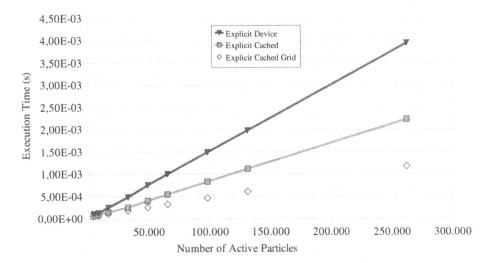

Fig. 8. Execution times for boxes of different sizes using the three explicit methods

Examining Fig. 7 in more detail, it becomes clear that the implicit method in CUDA does in fact perform well on the box dataset and only dissatisfactory on the heart dataset. Using OpenGL on the other hand, the implicit method performs comparably on both dataset. The reason is that the implicit method in CUDA cannot eliminate computation for inactive particles as effectively as the OpenGL implementation, which utilizes a hardware accelerated mask. For the box dataset this does not show since it contains no inactive particles.

Figure 8 shows that the execution times scales linearly with increasing input size (more than 50.000 active particles). For small datasets this may not be the case since small datasets do not utilize the parallel powers of the GPU fully.

From the discussion above it is evident that CUDA is a very interesting new platform to compute MSDMs in a surgical simulator. This requires however that additional aspects of a simulator, such as visualization and haptic feedback, are also adapted to the GPU. Fortunately, several techniques to implement such functionality are already described [6] [12] [13].

References

1. Liu, A., Tendick, F., Cleary, K., Kaufmann, C.: A survey of surgical simulation: applications, technology, and education. Presence: Teleoper. Virtual Environ. 12(6), 599–614 (2003)
2. Miller, K., Joldes, G., Lance, D., Wittek, A.: Total lagrangian explicit dynamics finite element algorithm for computing soft tissue deformation. Communications in Numerical Methods in Engineering 23, 121–134 (2007)
3. Irving, G., Teran, J., Fedkiw, R.: Invertible finite elements for robust simulation of large deformation. In: Eurographics/ACM SIGGRAPH Symposium on Computer Animation, pp. 131–140 (2004)
4. Bro-Nielsen, M., Cotin, S.: Real-time volumetric deformable models for surgery simulation using finite elements and condensation 15, 57–66 (1996)
5. Delingette, H., Cotin, S., Ayache, N.: A hybrid elastic model allowing real-time cutting deformations and force feedback for surgery training and simulation. In: Proceedings of the Computer Animation, May 1999, pp. 70–81 (1999)
6. Sørensen, T.S., Mosegaard, J.: An introduction to gpu accelerated surgical simulation. In: Harders, M., Székely, G. (eds.) ISBMS 2006. LNCS, vol. 4072, pp. 93–104. Springer, Heidelberg (2006)
7. Mosegaard, J., Sørensen, T.S.: Gpu accelerated surgical simulators for complex morphology. Proceedings of Virtual Reality 323, 147–154 (2005)
8. Taylor, Z., Cheng, M., Ourselin, S.: High-speed nonlinerar finite element analysis for surgical simulation using graphics processing units. IEEE Transactions on Medical Imaging (in Press, 2008)
9. NVIDIA: CUDA Programming Guide v. 1.1
10. Jakobsen, T.: Advanced character physics. In: Game Developers Conference (2001)
11. Sørensen, T., Stawiakski, J., Mosegaard, J.: Virtual open heart surgery: Obtaining models suitable for surgical simulation. Stud Health Technol. Inform. 125, 445–447 (2007)
12. Mosegaard, J., Sørensen, T.S.: Real-time deformation of detailed geometry based on mappings to a less detailed physical simulation on the gpu. In: Proceedings of Eurographics Workshop on Virtual Environments, vol. 11, pp. 105–111. Eurographics Association (2005)
13. Sørensen, T.S., Mosegaard, J.: Haptic feedback for the GPU-based surgical simulator. Proceedings of Medicine Meets Virtual Reality 14. Studies in Health Technology and Informatics 119, 523–528 (2006)

An Integrated Approach for Reconstructing a Surface Model of the Proximal Femur from Sparse Input Data and a Multi-Level Point Distribution Model

Guoyan Zheng and Miguel A. González Ballester

MEM Research Center - ISTB, University of Bern, Stauffacherstrasse 78,
CH-3014, Bern, Switzerland
guoyan.zheng@ieee.org

Abstract. In this paper, we present an integrated approach using a multi-level point distribution model (ML-PDM) to reconstruct a patient-specific surface model of the proximal femur from intra-operatively available sparse data, which may consist of sparse point data or a limited number of calibrated fluoroscopic images. We conducted experiments on clinical datasets as well as on datasets from cadaveric bones. Our experimental results demonstrate promising accuracy of the present approach. Further extension to reconstructing a surface model from pre-operative biplanar X-ray radiographs is discussed.

Keywords: reconstruction, point distribution model, statistical shape analysis.

1 Introduction

A patient-specific surface model of the proximal femur plays an important role in planning and supporting various computer-assisted surgical procedures including total hip replacement, hip resurfacing, and osteotomy of the proximal femur. The common approach to derive 3D models of the proximal femur is to use imaging technique such as computed tomography (CT) or magnetic resonance imaging (MRI). However, the high logistic effort, the extra radiation associated with the CT-imaging, and the large quantity of data to be acquired and processed make them less functional. The alternative presented here is to reconstruct the surface model using sparse input data consisting of dozens of surface points (e.g. 50 points) or a limited number of calibrated fluoroscopic images (e.g. 2 to 4 images).

Constructing an accurate 3D surface model from sparse input data is a challenging task. Additionally, inherent to the navigation application is the high accuracy and robustness requirements. When surface reconstruction is used for the purpose of surgical guidance, it requires that the algorithm satisfies the following criteria: (a) accurate geometrical information about the underlying anatomical structure can be derived from the reconstructed surface model; (b) target reconstruction error of the reconstructed surface model should be in the range of surgical usability, which is typically in the area of 1.5 mm average error (2 to 3 mm worst case) [1]; (c) 95% success rate is normally required, when an appropriate initialization is given [1]; (d) minimal user interaction during data acquisition and algorithm execution is highly appreciated for a sterilized environment;

F. Bello, E. Edwards (Eds.): ISBMS 2008, LNCS 5104, pp. 59–68, 2008.

and (e) the algorithm should be robust to outlying data. In the present paper, we try to solve the problem in an accurate and robust way. At the heart of our approach lies the combination of sophisticated surface reconstruction techniques and a multi-level point distribution model (ML-PDM) of the target anatomical structure.

The paper is organized as follows. Section 1 reviews the related work. Section 2 presents the construction of the ML-PDM of the proximal femur. Section 3 describes the integrated approach combining our previous works on 3D-3D surface reconstruction [2, 3] and those on 2D-3D surface reconstruction [4, 5]. Experiments and results are presented in Section 5, followed by conclusions in Section 6.

2 Related Works

Statistical shape analysis [6, 7, 8] is an important tool for understanding anatomical structures from medical images. A statistical model givens an effective parameterization of the shape variations found in a collection of sample models of a given population. Model based approaches [9, 10, 11] are popular due to their ability to robustly represent objects. Intraoperative reconstruction of a patient-specific model from sparse input data can be potentially achieved through the use of a statistical model. Statistical model building consists of establishing legal variations of shape from a training population. A patient-specific model is then instantiated through fitting the statistical model to intraoperatively acquired data. Thus, the aim of the statistical instantiation is to extrapolate from sparse input data a complete and accurate anatomical representation. This is particularly interesting for minimally invasive surgery (MIS), largely due to the operating theater setup.

Several research groups have explored the methods for reconstruction a patient-specific model from a statistical model and sparse input data such as digitized points [2, 3, 12, 13, 14, 15], a limited number of calibrated X-ray images [4, 5, 16, 17, 18, 19, 20], or tracked ultrasound [21, 22, 23, 24]. Except the method presented by Yao and Taylor [17], which depends on a deformable 2D/3D registration between an appearance based statistical model [25] and a limited number of X-ray images, all other methods have their reliance on a point distribution model (PDM) in common. In Fleute and Lavallée [12], a statistical shape model of the distal femur was fitted to sparse input points by simultaneously optimizing both shape and pose parameters. Their technology has been incorporated into a system for computer-assisted anterior cruciate ligament surgery and preliminary results were published in [13]. Chan et al. [21, 22] used a similar algorithm, but optimized the shape and pose parameters separately. Tracked ultrasound was used as the input in their work to instantiate 3D surface models of the complete femur and pelvis from their associated statistical shape models. Following the seminal work of Blanz and Vetter for the synthesis of 3D faces using a morphable model [26], Rajamani et al. [14, 15, 23] incorporated a Mahalanobis prior for a robust and stable surface model instantiation. In our recent work [2, 3], we proposed to use the dense surface point distribution model (DS-PDM) and a reconstruction scheme combining statistical instantiation and regularized shape deformation for an accurate and robust reconstruction of a patient-specific surface model of the proximal femur from dozens of points. This reconstruction scheme has also been combined with a novel 2D-3D correspondence establishing algorithm [4] for reconstructing surface model of the proximal femur from a limited number of calibrated X-ray images [5].

3 Multi-Level Point Distribution Model Construction

The ML-PDM used in this paper was constructed from a training database consisting of 30 proximal femoral surfaces from above the lesser trochanter. In the coarsest level, a sequence of correspondence establishing methods presented in [27] was employed to optimally align surface models segmented from CT volume. It started with a SPHARM-based parametric surface description [28] and then was optimized using Minimum Description Length (MDL) based principle as proposed in [29].

Following the alignment, the PDM in this level is constructed as follows. Let $\mathbf{x}_i = (p_0, p_1, ..., p_{N-1})$, $i = 0, 1, ..., m\text{-}1$, be m (here $m = 30$) members of the aligned training surfaces. Each member is described by a vectors $\mathbf{x}_i$ with N ($N = 4098$) vertices:

$$\mathbf{x}_i = \{x_0, y_0, z_0, x_1, y_1, z_1, ..., x_{N-1}, y_{N-1}, z_{N-1}\} \tag{1}$$

The PDM is obtained by applying principal component analysis on these surfaces.

$$D = ((m-1)^{-1}) \cdot \sum_{i=0}^{m-1}(\mathbf{x}_i - \overline{\mathbf{x}})(\mathbf{x}_i - \overline{\mathbf{x}})^T$$

$$P = (\mathbf{p}_0, \mathbf{p}_1, ...); \qquad D \cdot \mathbf{p}_i = \sigma_i^2 \cdot \mathbf{p}_i \tag{2}$$

Where $\overline{\mathbf{x}}$ and D represents the mean vector and the covariance matrix respectively. Then, any one of the instance in this space can be expressed as:

$$\mathbf{x} = \overline{\mathbf{x}} + \sum_{i=0}^{m-2} \alpha_i \mathbf{p}_i \tag{3}$$

And the estimated normal distribution of the coefficients $\{\alpha_i\}$ is:

$$p(\alpha_0, \alpha_1, ..., \alpha_{m-2}) = (2\pi)^{-\frac{m-1}{2}} \cdot e^{-\frac{1}{2}\sum_{i=0}^{m-2}(\alpha_i^2/\sigma_i^2)} \tag{4}$$

Where $\sum_{i=0}^{m-2}(\alpha_i^2/\sigma_i^2)$ is the Mahalanobis distance defined on the distribution.

The vertices for constructing the denser point distribution model in a finer resolution are then obtained by iteratively subdividing the aligned surface model in the coarser resolution. The basic idea of subdivision is to provide a smooth limit surface model which approximates the input data. Starting from a mesh in a low resolution, the limit surface model is approached by recursively tessellating the mesh. The positions of vertices created by tessellation are computed using a weighted stencil of local vertices. The complexity of the subdivision surface model can be increased until it satisfies the user's requirement.

In this work, we use a simple subdivision scheme called *Loop scheme*, invented by Loop [30], which is based on a spline basis function, called the three-dimensional quartic box spline. The subdivision principle of this scheme is very simple. Three new vertices are inserted to divide a triangle in a coarse resolution to four smaller triangles in a fine resolution.

Loop subdivision does not change the positions of vertices on the input meshes. Furthermore, positions of the inserted vertices in a fine resolution are interpolated from the neighboring vertices in a coarse resolution. As the input surface models have been optimized for establishing correspondences, it is reasonable to conclude that the output models are also aligned. Principal component analysis can be applied on these dense surface models to establish a dense surface point distribution model (DS-PDM). In our previous work [2], we found that a single level subdivision is enough for our purpose. We thus created a two-level point distribution model (TL-PDM).

4   The Integrated Surface Model Reconstruction Approach

Based on the two-level point distribution model, we developed an integrated surface model reconstruction approach which can seamlessly handle both 3D sparse points and a limited number of X-ray images. When a set of 3D points are used, the fine level point distribution model (FL-PDM) will be chosen, which facilitates the point-to-surface correspondence establishment. But if the input is a limited number of calibrated X-ray images, we will use the coarse level point distribution model (CL-PDM) to speed up the computation. For completeness, we will briefly present these two methods below. Details can be found in our previously published works [2, 3, 4, 5].

3D-3D Reconstruction Method [2, 3]: The reconstruction problem is formulated as a three-stage optimal estimation process. The first stage, *affine registration*, is to iteratively estimate the scale and the 6 degree-of-freedom rigid transformation between the mean shape of the PDM and the sparse input data using a correspondence building algorithm and a variant of iterative closest point (ICP) algorithm [31]. The estimation results of the first stage are used to establish point correspondences for the second stage, *statistical instantiation*, which optimally and robustly instantiates a surface model from the PDM using a statistical approach [14]. The instantiated surface model is taken as the input for the third stage, *regularized shape deformation*, where the input surface is further deformed by an approximating thin-plate spline (TPS) based vector transform [32] to refine the statistically instantiated surface model.

2D-3D Reconstruction Method [4, 5]: Our 2D-3D reconstruction approach combines statistical instantiation and regularized shape deformation as described above with an iterative image-to-model correspondence establishing algorithm [4]. The image-to-model correspondence is established using a non-rigid 2D point matching process, which iteratively uses a symmetric injective nearest-neighbor mapping operator and 2D thin-plate splines based deformation to find a fraction of best matched 2D point pairs between features detected from the calibrated X-ray images and the projections of the apparent contours extracted from the 3D model. The obtained 2D point pairs are then used to set up a set of 3D point pairs such that we turn a 2D-3D reconstruction problem to a 3D-3D one, which can be solved by the 3D-3D reconstruction approach as described above.

5   Experimental Results

We conducted experiments on 3 patient femurs and 18 cadaveric femurs (Note: none of them has been included for constructing the TL-PDM) with different shapes to validate the present approach. For each patient, we used two calibrated C-arm images acquired from the proximal femoral head as the input to our 2D-3D reconstruction scheme (the acquisition angle between these two images < 30°). Due to the lack of ground truth, we used the patient datasets to demonstrate qualitatively the 2D-3D reconstruction accuracy. One of the reconstruction examples is presented in Fig. 2. For the cadaveric femurs, 3D point sets as well as calibrated X-ray images were used.

Reconstruction error measurement: To quantify the reconstruction error in the cadaveric femur experiment, Target Reconstruction Error (TRE) was used. The TRE is defined as the distance between the actual and the reconstructed position of selected target features, which can be landmark points or bone surfaces themselves.

Validation experiments: Two different types of sparse data were explored in two experiments: (1) using clinically relevant sparse points directly acquired from the surfaces of 7 cadaveric femurs; (2) using a limited number of calibrated C-arm images of the other 11 cadaveric femurs. To evaluate the reconstruction error, we acquired another set of landmarks from each surface of the cadaveric femurs (Please note that these landmarks were not used in the reconstruction procedures but for the accuracy evaluation purpose). Then, the TRE was measured by calculating the distances between these landmarks and the associated reconstructed surface model.

Results of the first experiment: Total hip replacement and hip resurfacing procedures operated with posterior approach were identified as the potential clinical applications. At one stage of such surgeries, after internal rotation and posterior dislocation of the hip, most of the femoral head, neck, and some part of trochantric and intertrochantric (crest and line) regions are exposed [33]. Obtaining sparse surface points from these intraoperatively accessible regions and reconstructing a patient-specific 3D surface model of the proximal femur with reasonable accuracy will be useful for the above mentioned surgeries. In this experiment, one set of 50 points was used to reconstruct the surface model of each cadaveric bone and the other set consisted of 200 points was used to evaluate the reconstruction errors. The results of surface reconstruction using clinically relevant sparse points are presented in Fig. 1. For each case, the overall execution time was less than one minute

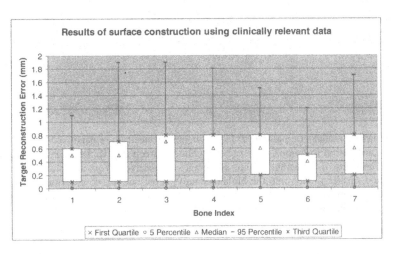

Fig. 1. Errors of reconstructing 3D surface models of seven cadaver femurs using clinically relevant 3D sparse point data

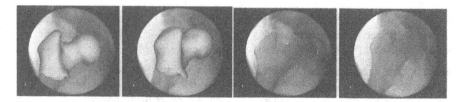

Fig. 2. An example of reconstructing a 3D surface model of the proximal femur from two calibrated C-arm images. Left: superposition of the reconstructed model on top of the images; right: the apparent contours extracted from the reconstructed model (yellow) vs. image contours (blue). An accurate match between the apparent contours and the input images is observed, even though part of the image contours (around greater trochanter) is mistakenly extracted.

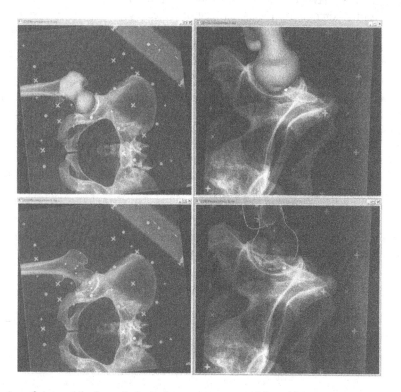

Fig. 3. A surface model of a pathologic femur reconstructed from biplanar radiographs by our 2D-3D reconstruction technique. An exact matching between the reconstructed surface model and the projections of the femur in the radiographs is observed. Top: superimposition of the reconstructed surface model onto the biplanar radiographs; bottom: comparison of the projection of the apparent contours (yellow) of the reconstructed surface model with the contours of the proximal femur (white). Our approach can achieve an accurate reconstruction even though certain part of the proximal femur is outside of the axial view (see the right images).

Results of the second experiment: In this experiment, two studies using different number of images were performed for each bone. In the first study two images acquired from anterior-posterior (AP) and lateral-medial (LM) directions were used to reconstruct the surface model of each cadaveric femur. In the second one, an image acquired from oblique direction was additionally used together with the above mentioned AP and LM images.

Table 1. Reconstruction errors when different number of images were used

Reconstruction errors when only AP and LM images were used for each bone										
Bone Index	1	2	3	4	5	6	7	8	9	10	11
Median (mm)	1.3	0.8	1.5	1	1.3	1	1.1	1	0.8	1.1	1.2
Mean (mm)	1.5	0.8	1.4	1.3	1.4	1.2	1.2	1.2	1	1.1	1.6
Reconstruction errors when all three images were used for each bone										
Bone Index	1	2	3	4	5	6	7	8	9	10	11
Median (mm)	1.3	0.7	0.7	1.1	1	1.1	0.8	0.9	0.7	1	0.9
Mean (mm)	1.3	0.7	0.8	1.2	1.1	1.1	1.1	0.9	0.9	1.1	1.2

The reconstruction accuracies were evaluated by randomly digitizing 100 – 200 points from each surface of the cadaveric specimen and then computing the distance from those digitized points to the associated surface reconstructed from the images. The median and mean reconstruction errors of both experiments are presented in Table I. An average mean reconstruction error of 1.2 mm was found when only AP and LM images were used for each bone. It decreased to 1.0 mm when three images were used.

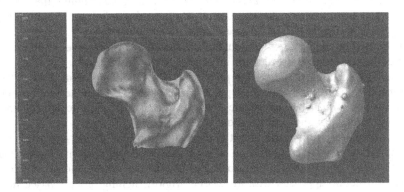

Fig. 4. Color-coded error TRE distribution when the reconstructed surface model as shown in Fig. 3 was compared to its ground truth obtained by a laser-scan reconstruction method. The maximum TRE is 3.7 mm and the mean TRE is 0.7 mm. Note that we only computed the TREs from the reconstructed surface model to the ground truth as the former was smaller than the latter.

6   Conclusions

We have presented an integrated approach using the ML-PDM for robust and accurate anatomical shape reconstruction from sparse input data. Based on the modalities of the input data, the appropriate level point distribution model was used. In this approach, the 3D-3D reconstruction problem is formulated as a three-stage optimal estimation process. In each stage, the best result is optimally estimated under the assumption for that stage, which guarantees a topologically preserved solution when only sparse points are available. The FL-PDM is employed in all stages to facilitate the correspondence establishment. When a limited number of calibrated X-ray images are used, the CL-PDM is employed to speed up the computation. A 2D-3D correspondence establishing algorithm based on a non-rigid 2D point matching process is applied to convert a 2D-3D problem to a 3D-3D one.

The reason why we require a ML-PDM can be summarized as follows. Due to the sparseness of the point data, we need a FL-PDM to establish precise correspondences between each input point and the point on the FL-PDM in the second stage of our 3D-3D reconstruction method, although it is true that this is not an important issue in the first and the third stage of our 3D-3D reconstruction approach. In these two stages we even can use the CL-PDM together with a KD-Tree data structure to accelerate the correspondence establishment procedure. However, if a CL-PDM is used in the second stage, each input point will find a point on the surface model described by the CL-PDM as its corresponding point. However, this point is not necessary a vertex of the CL-PDM. Then to solve the statistical instantiation problem, we either need to use the closest vertex as the correspondence of the input point, or we need to do an interpolation. The former method will bring the correspondence error into the statistical instantiation procedure due to the relative sparseness of the CL-PDM, whereas the latter method will complicate the solution to the statistical instantiation problem. On contrast, this is different when the co-registered images are taken as the input for the 2D-3D reconstruction. Although the number of input images is limited, the extracted edge contours from these images are dense (a sub-pixel resolution). In such a situation, we don't need a FL-PDM to set up a precise correspondence. If a vertex on the PDM is located on the apparent contours of the model and is preserved even after the 2D-3D correspondence establishment procedure, we can always find a 3D point on the forward projection ray of an edge point as its corresponding point. Thus, at each step of the 2D-3D reconstruction, new points on both the PDM and the projection rays of the edge points may be changeable, while in the 3D-3D case the input point set remains the same throughout all iteration.

Furthermore, we are investigating the reconstruction of a patient-specific surface model of the proximal femur from a pair of calibrated pre-operative X-ray radiographs [34, 35]. The results are encouraging. Due to the combination of our three-stage reconstruction approach with the 2D non-rigid matching process based image-to-model correspondence establishing algorithm, our method can handle both non-pathological and pathological cases even when the ML-PDM is constructed from the surface models of normal patients, which is an important advantage over the existing works. An example of reconstructing a surface model of a pathological femur from a pair of the calibrated X-ray radiographs and the distribution of the TREs when the reconstructed surface model is compared to its ground truth (a surface model extracted by a laser-scan with the precision of 0.1 mm) is represented in Fig. 3 and Fig. 4, respectively. We have adapted the open-source software MESH [36] to compute the distribution of the TREs along the whole surface of the reconstructed model. An average TRE of 0.9 mm was found when 22 femurs with both non-pathologic and pathologic bones were considered. For details about the calibration of the X-ray radiograph as well as the reconstruction method, we refer to [34] and [35].

The proposed approach is generic and can be easily extended to other rigid anatomical structures, though in this paper we only demonstrate its application for reconstructing surface models of the proximal femur.

References

1. Livyatan, H., Yaniv., Z., Joskowicz, Z.: Gradient-based 2-D/3-D rigid registration of fluoroscopic X-ray to CT. IEEE T. Med. Imaging 22(11), 1395–1406 (2003)
2. Zheng, G., Rajamani, K.T., Nolte, L.-P.: Use of a dense point distribution model in a three-stage anatomical shape reconstruction from sparse information for computer assisted orthopaedic surgery: a preliminary study. In: Narayanan, P.J., Nayar, S.K., Shum, H.-Y. (eds.) ACCV 2006. LNCS, vol. 3852, pp. 52–60. Springer, Heidelberg (2006)
3. Zheng, G., Dong, X., Nolte, L.-P.: Robust and accurate reconstruction of patient-specific 3D surface models from sparse point sets: a sequential three-stage trimmed optimization approach. In: Yang, G.-Z., Jiang, T., Shen, D., Gu, L., Yang, J. (eds.) MIAR 2006. LNCS, vol. 4091, pp. 68–75. Springer, Heidelberg (2006)
4. Zheng, G.: A novel 3D/2D correspondence building method for anatomy-based registration. In: Pluim, J.P.W., Likar, B., Gerritsen, F.A. (eds.) WBIR 2006. LNCS, vol. 4057, pp. 75–83. Springer, Heidelberg (2006)
5. Zheng, G., et al.: Reconstruction of patient-specific 3D bone surface from 2D calibrated fluoroscopic images and point distribution model. In: Larsen, R., Nielsen, M., Sporring, J. (eds.) MICCAI 2006. LNCS, vol. 4190, pp. 25–32. Springer, Heidelberg (2006)
6. Dryden, I.L., Mardia, K.V.: Statistical Shape Analysis. John Wiley, Chichester (1998)
7. Kendall, D.: A survey of the statistical theory of shape. Statistical Science 4(2), 87–99 (1989)
8. Small, C.: The statistical Theory of Shape. Springer, Heidelberg (1996)
9. Turk, M., Pentland, A.: Eigenfaces for recognition. Journal of Cognitive Neuroscience 3(1), 71–86 (1991)
10. Cootes, T.F., et al.: Active shape models – their training and application. Computer Vision and Image Understanding 61(1), 38–59 (1995)
11. Corouge, I., et al.: Interindividual functional mapping: a nonlinear local approach. Neuroimage 19, 1337–1348 (2003)
12. Fleute, M., Lavallée, S.: Building a complete surface model from sparse data using statistical shape models: application to computer assisted knee surgery system. In: Wells, W.M., Colchester, A.C.F., Delp, S.L. (eds.) MICCAI 1998. LNCS, vol. 1496, pp. 879–887. Springer, Heidelberg (1998)
13. Fleute, M., Lavallée, S., Julliard, R.: Incorporating a statistically based shape model into a system for computer-assisted anterior cruciate ligament surgery. Medical Image Analysis 3(3), 209–222 (1999)
14. Rajamani, K.T., Styner, M., Joshi, S.C.: Bone model morphing for enhanced surgical visualization. Proceedings of the ISBI 2004, 1255–1258 (2004)
15. Rajamani, K.T., et al.: A novel and stable approach to anatomical structure morphing for enhanced intraoperative 3D visualization. In: Proceedings of the SPIE Medical Imaging: Visualization, Image-guided Procedures, and Display, vol. 5744, pp. 718–725 (2005)
16. Fleute, M., Lavallée, S.: Nonrigid 3-D/2-D registration of images using statistical models. In: Taylor, C., Colchester, A. (eds.) MICCAI 1999. LNCS, vol. 1679, pp. 138–147. Springer, Heidelberg (1999)

17. Yao, J., Taylor, R.H.: Assessing accuracy factors in deformable 2D/3D medical image registration using a statistical pelvis model. In: Proceedings of ICCV 2003, vol. 2, pp. 1329–1334. IEEE Computer Society, Los Alamitos (2003)

18. Lamecker, H., Wenckebach, T.H., Hege, H.-C.: Atlas-based 3D-shape reconstruction from X-ray images. In: Proceedings of ICPR 2006, vol. 1, pp. 371–374 (2006)

19. Benameur, S., et al.: 3D/2D registration and segmentation of scoliotic vertebrae using statistical models. Computerized Medical Imaging and Graphics 27, 321–337 (2003)

20. Benameur, S., et al.: A hierarchical statistical modeling approach for the unsupervised 3-D biplanar reconstruction of the scoliotic spine. IEEE Transactions on Biomedical Engineering 52(12), 2041–2057 (2005)

21. Chan, C.S.K., Edwards, P.J., Hawkes, D.J.: Integration of ultrasound-based registration with statistical shape models for computer-assisted orthopaedic surgery. In: Proceedings of SPIE Medical Imaging 2003: Image Processing, vol. 5032, pp. 414–424 (2003)

22. Chan, C.S.K., Barratt, D.C., Edwards, P.J., Penney, G.P., Slomczykowski, M., Carter, T.J., Hawkes, D.J.: Cadaver validation of the use of ultrasound for 3D model instantiation of bony anatomy in image guided orthopaedic surgery. In: Barillot, C., Haynor, D.R., Hellier, P. (eds.) MICCAI 2004. LNCS, vol. 3217, pp. 397–404. Springer, Heidelberg (2004)

23. Talib, H., Rajamani, K.T., Kowal, J., Nolte, L.-P., Styner, M., Gonzalez Ballester, M.A.: A comparison study assessing the feasibility of ultrasound-initialized deformable bone models. Computer Aided Surgery 10(5/6), 293–299 (2005)

24. Talib, H., et al.: Feasibility of 3D ultrasound-initialized deformable bone-modeling. In: Proceedings of CAOS 2006, pp. 519–522 (2006)

25. Yao, J., Taylor, R.H.: Tetrahedral mesh modeling of density data for anatomical atlases and intensity-based registration. In: Delp, S.L., DiGoia, A.M., Jaramaz, B. (eds.) MICCAI 2000. LNCS, vol. 1935, pp. 531–540. Springer, Berlin (2000)

26. Blanz, V., Vetter, T.: A morphable model for the synthesis of 3D faces. In: Proceedings of the 26th Annual Conference on Computer Graphics, SIGGRAPH 1999, pp. 187–194 (1999)

27. Styner, M., et al.: Evaluation of 3D correspondence methods for modeling building. In: Taylor, C.J., Noble, J.A. (eds.) IPMI 2003. LNCS, vol. 2732, pp. 63–75. Springer, Heidelberg (2003)

28. Brechbuehler, C., Gerig, G., Kuebler, O.: Parameterization of Closed Surfaces for 3D Shape Description. Comput Vision and Image Under 61, 154–170 (1995)

29. Davies, R.H., Twining, C.H., et al.: 3D statistical shape models using direct optimization of description length. In: Heyden, A., Sparr, G., Nielsen, M., Johansen, P. (eds.) ECCV 2002. LNCS, vol. 2352, pp. 3–20. Springer, Heidelberg (2002)

30. Loop, C.T.: Smooth subdivision surfaces based on triangles. M.S. Thesis, Department of Mathematics, University of Utah (August 1987)

31. Besl, P., McKay, N.D.: A method for registration of 3D shapes. IEEE Transaction on Pattern Analysis and Machine Intelligence 14(2), 239–256 (1992)

32. Bookstein, F.: Principal warps: thin-plate splines and the decomposition of deformations. IEEE Transaction on Pattern Analysis and Machine Intelligence 11(6), 567–585 (1989)

33. Moreland, J.R.: Primary Total Hip Arthroplasty. In: Chapman, M.W. (ed.) Operative Orthopaedics, 1st edn., vol. 1, pp. 679–693. J.B Lippincott, Philadelphia (1988)

34. Schumann, S., Zheng, G., Nolte, L.-P.: Calibration of X-ray radiographs and its feasible application for 2D/3D reconstruction of the proximal femur. IEEE EMBS (submitted, 2008)

35. Zheng, G., Schumann, S.: 3-D reconstruction of a surface model of the proximal femur from digital biplanar radiographs. IEEE EMBS (submitted, 2008)

36. Aspert, N., Santa-Cruz, D., Ebrahimi, T.: MESH: measuring error between surface using the Hausdorff distance. In: Proceedings of the IEEE International Conference on Multimedia and Expo 2002 (ICME), vol. I, pp. 705–708 (2002)

Coupling Deformable Models
for Multi-object Segmentation

Dagmar Kainmueller, Hans Lamecker, Stefan Zachow, and Hans-Christian Hege

Zuse Institute Berlin, Germany

Abstract. For biomechanical simulations, the segmentation of multiple adjacent anatomical structures from medical image data is often required. If adjacent structures are hardly distinguishable in image data, automatic segmentation methods for single structures in general do not yield sufficiently accurate results. To improve segmentation accuracy in these cases, knowledge about adjacent structures must be exploited. Optimal graph searching based on deformable surface models allows for a simultaneous segmentation of multiple adjacent objects. However, this method requires a correspondence relation between vertices of adjacent surface meshes. Line segments, each containing two corresponding vertices, may then serve as shared displacement directions in the segmentation process. The problem is how to define suitable correspondences on arbitrary surfaces. In this paper we propose a scheme for constructing a correspondence relation in adjacent regions of two arbitrary surfaces. When applying the thus generated shared displacement directions in segmentation with deformable surfaces, overlap of the surfaces is guaranteed not to occur. We show correspondence relations for regions on a femoral head and acetabulum and other adjacent structures, as well as preliminary segmentation results obtained by a graph cut algorithm.

1   Introduction

For patient-specific biomechanical simulations, e.g. of the human lower limb, an accurate reconstruction of the bony anatomy from medical image data is required. This particularly applies to joint regions, as simulation results heavily depend on the location of joints. In CT data, bony tissue can usually be reconstructed by simple thresholding. However, in joint regions, thresholding is often not sufficient for separating adjacent individual bones from each other. Due to large slice distances or pathological changes of the bones, the joint space may be hard to detect even for a human observer. Fig. 1a and 1b show exemplary situations.

We can achieve good initializations for individual (single) bony structures (e.g. pelvis, femur) by our segmentation framework which is based on statistical shape models (SSM) and free form models (FFM) [1]. Segmentations with SSMs yield good initializations, but lack precision, since previously unknown patient-specific anatomy is generally not contained in the model. With FFMs more precise segmentations can be achieved, but they suffer from a loss of shape

F. Bello, E. Edwards (Eds.): ISBMS 2008, LNCS 5104, pp. 69–78, 2008.
© Springer-Verlag Berlin Heidelberg 2008

knowledge, causing inaccurate interpolations where the object to be segmented cannot be distinguished from adjacent structures in image data. Furthermore, the lack of image information may generally lead to overlapping segmentation results when adjacent structures are segmented separately.

A basic idea for improving segmentation results and simultaneously solving the overlap problem is to segment multiple adjacent objects at the same time and incorporate some knowledge about their spatial relationship. The problem we are concerned with in this paper is how to establish a suitable coupling of two arbitrary adjacent deformable surface models (triangular meshes), assuming that a good initialization of the two models is given. The contribution of this work is a construction scheme for shared displacement directions for two arbitrary surfaces, that is, line segments along which vertices of both surfaces can be displaced in a deformable surface segmentation framework. We used the coupling realized by these shared displacement directions for fine grain multi-object segmentation based on graph cuts. This paper presents a proof of concept with very encouraging results.

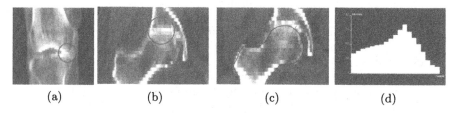

(a)   (b)   (c)   (d)

Fig. 1. (a) CT data of distal femur and proximal tibia, slice distance 2mm, (b) acetabulum and proximal femur, slice distance 4,6mm. Joint space hardly visible in encircled areas. (c) Acetabulum and proximal femur, slice distance 5mm, with surface model cross-section (black) and domain of intensity profile (red). (d) Intensity profile for domain in (c).

2   Related Work

Costa et al. [2] employ a non-overlapping constraint for coupled segmentation of prostate and bladder with deformable models. They propose a force that drives two models apart if intersections occur in the segmentation process. This method principally allows for free form deformations of the models while coping with overlap. If present, a statistical shape model can be enforced on one of the models. If one structure is better distinguishable from the background than the other, an asymmetric non-overlap force can be applied. This approach yields promising results in prostate and bladder segmentation. However, displacements are not found *simultaneously* for both objects: Apart from the non-overlap force which only exists if an overlap has already occurred, one object does not take into account any knowledge about the presence of the other object in displacement computation.

Tsai et al. [3] build composite statistical shape models by applying principal component analysis to a training set of implicit (signed distance function) representations of multiple objects. They apply such models to the segmentation of subcortical brain structures and male lower abdominal structures (prostate gland, rectum, obturator muscles). Babalola et al. [4] build a composite active appearance model on the basis of explicit (surface-mesh) representations of multiple subcortical brain structures. They apply this model to obtain a good initialization of brain structure models to accurately segment the caudate in a single object segmentation framework. Composite statistical shape models yield a tight coupling of the deformations of multiple objects. Ideally, no overlap between adjacent objects should be possible in model space. Anyway, neither of the two approaches allows for a fine grain free form multi-object segmentation, as model deformation is bound to the respective shape space.

Li et al. [5] solve the overlap problem with optimal graph searching in a deformable model segmentation framework. They apply their method to the segmentation of bone and cartilage in 3D MRI of human ankles. Vertex normals of the bone surface are used as shared displacement directions for bone and cartilage surfaces. Thus, one direction may be used to search for two object boundaries, thereby allowing for a completely simultaneous segmentation process. However, methods involving shared displacement directions have been described for surfaces on which corresponding vertices are easily found. This holds for height field or cylindrical surfaces in regular grids [6], or if one surface can be obtained by displacing the other along its vertex normals [5]. To the best of our knowledge, there is at present no way for generating shared displacement directions on arbitrary surfaces.

3   Multi-object Segmentation with Graph Cuts

In segmentation with deformable surfaces, *intensity profiles* are commonly used to guide the deformation process. Intensity profiles are intensities sampled f image data along line segments at each vertex of the surface mesh, see Fig. 1c and 1d. Note that the term *intensity profile* or just *profile* may be used as referring to the sampled intensities only, but here we use it to refer to the domain of the sampling, i.e. a line segment, as well. Fig. 2a shows exemplary profiles on a triangular surface mesh. Profiles are commonly defined to run along vertex normals, but other directions may be chosen as well. On each profile, a cost function is derived from image data for a number of equidistant sampling points. The minimum cost sample point on a profile may serve as a desired (locally optimal) position for the respective vertex.

However, graph cut algorithms allow for a *global* optimization of the sum of costs for each vertex displacement while respecting hard constraints on the distance of multiple objects and on single object smoothness. The hard constraints are realized by means of graph edges that connect sample points on profiles in such a way that a non empty *minimum closed set* in the graph defines the optimal surfaces. A smoothing constraint guarantees that new vertex positions on adjacent profiles are at most s sample points away from each other. If sample point i is chosen on a profile as desired new position, sample point i-s or higher must be chosen on adjacent profiles, as illustrated in Fig. 2b. The smaller s is chosen, the

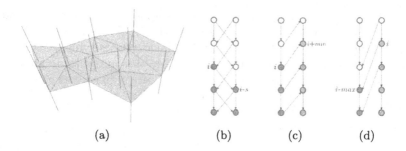

Fig. 2. (a) Triangular surface mesh (red) with profiles (black). (b) Graph edges realizing smoothing constraint. Here, $s = 1$. (c) Minimum distance constraint on shared profile: Sample point i for surface A (blue) entails sample point $i+min$ or higher for surface B (red) on duplicated profile. Here, $min = 1$. (d) Maximum distance constraint: Sample point i for surface B entails sample point $i-max$ or higher for surface A. Here, $max = 2$.

smoother is the surface resulting from the optimization. Multiple surfaces can be coupled with shared intensity profiles at individual vertices. Shared profiles are duplicated in the graph, so that each surface is equipped with an instance of each shared profile. Minimum and maximum distance constraints guarantee that new vertex positions found on shared profiles by graph optimization are at least min and at most max sample points away from each other, see Fig. 2c and 2d. For more details on graph construction see [6]. The minimum closed set problem can be transformed to a minimum s-t-cut problem, which is solved in polynomial time by maximum flow algorithms [7].

4   Non-overlapping Surface Deformations

Before shared profiles can be constructed, surface models of adjacent structures must be initialized to have a reasonable spatial relation to each other. For bony structures of the lower limb we can achieve initializations in CT data with a maximum surface distance of about 1cm to a manual expert segmentation. For two such well-initialized surface models, we identify a potential overlap region by growing each surface by a user-specified profile length, see Fig. 3a and 3b. Our concern is to couple the surfaces with shared profiles wherever single profiles would reach the overlap region.

4.1   Properties of Shared Intensity Profiles

Which conditions must hold on shared profiles to generally prevent any overlap after surface deformation? Note one necessary condition posed on new vertex positions of coupled surface meshes A and B: Let x be a vertex on A and $m(x)$ the corresponding vertex on B. Then the new position x_A^* of x must lie closer to A than the new position x_B^* of $m(x)$ in profile direction, i.e. $\|x_A^* - x\| < \|x_B^* - x\|$ (if we disregard that surfaces can completely swap sides along shared profiles). See Fig. 3c. In the following we assume that this condition holds.

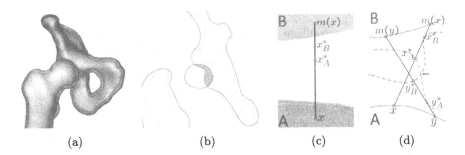

(a)   (b)   (c)   (d)

Fig. 3. (a) Proximal femur and ilium with transparent isosurfaces at same distance. (b) 2D cross-section. Thin lines: surface contours. Thick lines: isosurface contours at the same distance. Grey region: potential overlap region. (c) Necessary condition: x_A^* lies closer to x than x_B^*. (d) Contours A (blue) and B (red). Black lines: Intersecting connections of mapped points. Dotted lines: Exemplary deformed contours with intersection (black arrow).

In 2D, when establishing shared profiles between deformable contours A and B, intersecting profiles can cause overlap of the deformed contours, as illustrated in Fig. 3d. Now we examine this situation in 3D. For this purpose we consider a bijective mapping m of piecewise *continuous* regions R_A and R_B on surfaces A and B: Shared profiles on triangular meshes can be seen as a finite set of line segments connecting corresponding triangle vertices x and $m(x)$. Hence they partially define or are embedded in such a mapping. However, during surface deformation, not only each vertex, but *each point* on R_A and R_B is displaced along a line segment that leads to its corresponding point. Profiles may not intersect with each other, while other line segments that connect mapped point pairs do. Thus we are interested in properties of mappings m and not only in properties of shared profiles.

Overlap after surface deformation cannot occur if the mapping m satisfies what we call *non-intersection* condition, i.e. no two connections of two mapped point pairs intersect. For a proof, let R_A^* and R_B^* be the deformed regions. R_A^* is the image of a function f_A that maps each point x on R_A to a point x_A^* on the line segment $\{x + \lambda \cdot (m(x) - x) | 0 \le \lambda \le 1\}$. Each x_A^* is defined by an individual $\lambda_A(x) \in [0,1]$. Likewise R_B^* is the image of f_B that maps each x on R_A to an $x_B^* = x + \lambda_B(x) \cdot (m(x) - x)$, with the additional constraint that $\lambda_A(x) < \lambda_B(x)$ for all $x \in R_A$.

If the deformed regions intersect, we have $x, y \in R_A$ with $f_A(x) = f_B(x)$, i.e. $x + \lambda_A(x) \cdot (m(x) - x) = y + \lambda_B(y) \cdot (m(y) - y)$. The constraint $\lambda_A(x) < \lambda_B(x)$ implies $x \ne y$. This implies an intersection of the line segments $[x, m(x)]$ and $[y, m(y)]$. In reverse, if no two connections of two mapped point pairs intersect, the deformed regions do not intersect either. Note that the non-intersection condition also implies the continuity of m for homeomorphic regions R_A and R_B, yet continuity alone is not a sufficient criterion for preventing overlap.

4.2   Mapping of Non-empty Surface Regions

The following scheme establishes a mapping of topologically equivalent regions on smooth, closed surfaces A and B that satisfies the non-intersection condition:

1. Compute the mid surface M as all points with the same distance to A and B. Compute the normals n on M, wherever M is smooth.
2. Define finite length vectors v on M as follows: Scale normals n and inverse normals $-n$ to some length exceeding the maximum distance of any two points x on A and y on B. Then trim the scaled normals at the skeleton of M if they reach it. (The skeleton of a surface is the set of points that are centers of a sphere that touches the surface in more than one point, but does not cut it.) The resulting vectors then do not intersect with each other.
3. Iteratively map points on A and B cut by the same vector v. Start with vectors on local minima of the signed euclidean distance function $d : M \to \mathbb{R}, x \mapsto d(x, A)$ on M. Grow the regions considered on M as long as corresponding regions on A and B have the same topologies.

This results in an intersection free mapping of regions. The mapped regions are not empty: At least all normals on M where the distance function $d : M \to \mathbb{R}, x \mapsto d(x, A)$ has a local minimum connect corresponding points. (Note that for all x $d(x, A) = d(x, B)$.) Such local minima exist, as we are dealing with closed surfaces. The linear connections of the respective corresponding points also do not contain any point of the skeleton of M.

As a proof, we first examine properties of the normals on M at local minima of d: The connection of any point x on M to a closest point a_x on A is normal on A. (Imagine a sphere with radius $d(x)$ around x. This sphere touches A in a_x.) This applies to closest points b_x on B, accordingly. Local minima of the function d are located at points m on M where isosurfaces of the distance transforms of A and B *touch*. Let a_m and b_m be closest points to m on A and B, respectively. As isosurfaces with value $d(m)$ touch in m and the distances from a_m to m and b_m to m are both $d(m)$, spheres around a_m and b_m with radius $d(m)$ touch in m. This implies that the connections between a_m and m as well as b_m and m both are perpendicular to M. The direct connection between a_m and b_m therefore contains m and is perpendicular to A, B and M.

Now we show that the connection g between a_m and b_m does not meet the skeleton of M: Assume it did. Let p be the closest point to m on g that lies on the skeleton. Assume without loss of generality that p lies closer to a_m than to b_m. Then there is a circle around p that touches M in multiple points, but does not intersect with M. Let r denote the radius of that circle R. r must be smaller or equal to the distance $\|p - m\|$, otherwise m would lie *inside* this circle. Let n be one of the points that R touches, $n \neq m$. Then the distance between n and a_m is shorter than the distance between m and a_m. This is a contradiction to m being the closest point to a_m on M. Hence, there is no point on g that lies on the skeleton of M.

4.3   Generating Shared Intensity Profiles

Based on the scheme for constructing an intersection-free mapping as proposed in Section 4.2, we realize a construction algorithm for shared profiles on pairs of adjacent *triangular* surfaces A and B. In the process, we modify the connectivity of parts of the surface meshes. The one-sided surface distance of the modified surface to the original surface is always zero. Concerning the reverse direction of the surface distance, no general assertions can be made.

Fig. 4 shows the construction pipeline for an exemplary femur and ilium. First, the mid surface M between the objects is computed as the zero level of the objects' distance transforms, subtracted from each other (Fig. 4b). M is discretized and triangulated to form a mesh of high regularity. The skeleton of M is approximated by uniformly displacing M along its surface normals in both directions and identifying self intersection points, see Fig. 5. Then we identify the region on M where its vertex normals, scaled to a user-specified maximal length, enter both femur and ilium, without reaching the skeleton of M first (Fig. 4c and 4d). This region is then displaced onto the surfaces of each femur and ilium in vertex normal direction. These displaced patches are merged into the respective surface mesh by removing the original triangles in this region and connecting the boundary of the remaining mesh to the boundary of the patch. The resulting surfaces are shown in Fig. 4e and 4f.

As a result, we obtain a bijective mapping of the displaced (continuous) patches that satisfies the non-intersection condition. Thereby shared profiles between the modified surface meshes' vertices are defined, as shown in Fig. 4g and 4h. We let the shared profiles reach *into* the surfaces until they meet the skeleton of M, or the inner skeleton of the respective surface, or they reach

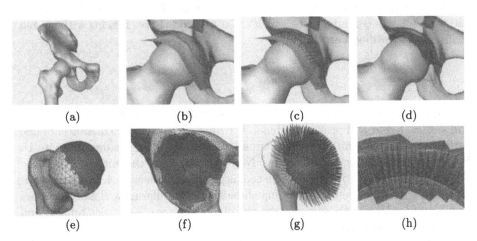

(a)   (b)   (c)   (d)

(e)   (f)   (g)   (h)

Fig. 4. Construction of shared profiles. (a) Proximal femur and right ilium. (b) Mid surface (yellow) in region of interest. (c) Normals on mid surface entering both femur and illium. (d) Extracted region on mid surface (red). (e,f) femoral head and acetabulum with integrated displaced region (red). (g,h) Vectors coupling femoral head and acetabulum (black).

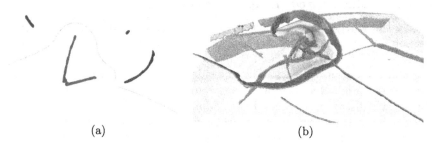

(a)   (b)

Fig. 5. 2D cross-section (a) and 3D view (b) of mid surface (yellow), upper skeleton (blue) and lower skeleton (red). For a better impression of the skeleton, the mid surface shown here is not restricted to a region of interest.

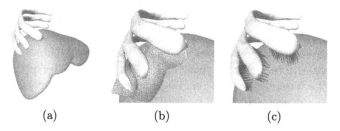

(a)   (b)   (c)

Fig. 6. (a) Surface models of liver (red) and nearby ribs (yellow). (b) Mid surface in region of interest. (c) Coupling vectors between liver and ribs.

a user-specified maximal length. As another example, Fig. 6 shows a liver model and a model of surrounding ribs which are connected. Three regions with shared profiles are identified here.

5   Results

In a first investigation we computed segmentations of a femoral head and ilium and a distal femur and proximal tibia in CT data with 5mm slice distance. Single surface models were initialized individually with SSM fitting as described in [1]. Fig. 7a shows resulting initializations. Note the overlap of the initialized models. Shared profiles (Fig. 7b) were established as proposed in Section 4.3. A graph was then constructed as in [6], containing these shared profiles as well as traditional single profiles [1] in non-adjacent regions of the surfaces. As a cost function on the profiles we used thresholding in the image data, see Fig. 7d. Costs are low at sample points where intensities change from above to below the threshold, and high everywhere else. We applied the graph cut algorithm proposed in [7] to find optimal surfaces. Fig. 7c shows the resulting segmentations. The overlap was resolved and a reasonable interpolation was found in image regions where intensities lie above the threshold due to low image quality.

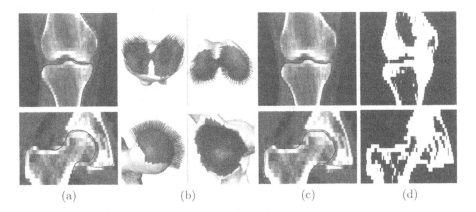

Fig. 7. First row: Distal femur and proximal tibia. Second row: Acetabulum and proximal femur. (a) Model initialization with statistical shape models. (b) Shared profiles. (c) Graph cuts optimization. (d) Threshold information used for model fitting.

6 Discussion

We proposed a method for coupling arbitrary adjacent surfaces by means of shared profiles. The shared profiles are constructed in a way that they satisfy a non-intersection condition. This property guarantees that two surfaces do not overlap if deformed along the shared profiles. We applied the coupling method to femur, tibia and ilium models which were initialized in exemplary CT data by SSM fitting. Optimal graph searching on the graph obtained from the coupled surfaces produced encouraging results.

The next task will be to evaluate segmentation results quantitatively on a set of 3D CT datasets by comparing them to manual expert segmentations. For this purpose, as a first step, variations among segmentations performed by different experts must be determined. This is of particular importance here as we are dealing with segmentations of adjacent objects which often are hardly distinguishable in image data, even for human observers. Until now, simple thresholding was performed for establishing a cost function. We are working on a more elaborate cost function design that considers more information (e.g. the gradient) contained in image data. We also plan to establish an automatic method for rough initialization of pelvis and femur models in CT datasets as starting positions for SSM segmentation. Until now, this first initialization requires manual interaction.

Acknowledgements

Dagmar Kainmueller is funded by DFG Collaborative Research Center SFB760. Hans Lamecker is funded by DFG Research Center Matheon in Berlin. Thanks to Heiko Seim (ZIB) for providing the shape models of femur and tibia. Thanks to Charité Center for Musculoskeletal Surgery for providing the pelvis model.

References

1. Kainmueller, D., Lange, T., Lamecker, H.: Shape Constrained Automatic Segmentation of the Liver based on a Heuristic Intensity Model. In: 3D Segmentation in the Clinic: A Grand Challenge, pp. 109–116 (2007)
2. Costa, M.J., Delingette, H., Novellas, S., Ayache, N.: Automatic segmentation of bladder and prostate using coupled 3d deformable models. In: Ayache, N., Ourselin, S., Maeder, A. (eds.) MICCAI 2007, Part I. LNCS, vol. 4791, pp. 252–260. Springer, Heidelberg (2007)
3. Tsai, A., Wells, W., Tempany, C., Grimson, W.E.L., Willsky, A.S.: Coupled multi-shape model and mutual information for medical image segmentation. In: Taylor, C.J., Noble, J.A. (eds.) IPMI 2003. LNCS, vol. 2732, pp. 185–197. Springer, Heidelberg (2003)
4. Babalola, K.O., Petrovic, V., Cootes, T., Taylor, C., Twining, C., Williams, T., Mills, A.: Automatic Segmentation of the Caudate Nuclei using Active Appearance Models. In: 3D Segmentation in the Clinic: A Grand Challenge, pp. 57–64 (2007)
5. Li, K., Millington, S., Wu, X., Chen, D.Z., Sonka, M.: Simultaneous segmentation of multiple closed surfaces using optimal graph searching. In: Christensen, G.E., Sonka, M. (eds.) IPMI 2005. LNCS, vol. 3565, pp. 406–417. Springer, Heidelberg (2005)
6. Li, K., Wu, X., Chen, D.Z., Sonka, M.: Optimal surface segmentation in volumetric images-a graph-theoretic approach. IEEE Trans. Pattern Anal. Mach. Intell. 28(1), 119–134 (2006)
7. Boykov, Y.Y., Kolmogorov, V.: An experimental comparison of min-cut/max-flow algorithms for energy minimization in vision. IEEE Trans. Pattern Analysis and Machine Intelligence 26(9), 1124–1137 (2004)

Modelling Prostate Gland Motion for Image-Guided Interventions

Yipeng Hu[1], Dominic Morgan[1], Hashim Uddin Ahmed[2], Doug Pendsé[3,4],
Mahua Sahu[3], Clare Allen[4], Mark Emberton[2], David Hawkes[1], and Dean Barratt[1]

[1] Centre for Medical Image Computing, Department of Medical Physics & Bioengineering;
[2] Department of Urology, Division of Surgery & Interventional Science; and
[3] National Medical Laser Centre, University College London, London, UK
[4] Department of Radiology, University College Hospital, London, UK

Abstract. A direct approach to using finite element analysis (FEA) to predict organ motion typically requires accurate boundary conditions, which can be difficult to measure during surgical interventions, and accurate estimates of soft-tissue properties, which vary significantly between patients. In this paper, we describe a method that combines FEA with a statistical approach to overcome these problems. We show how a patient-specific, statistical motion model (SMM) of the prostate gland, generated from FE simulations, can be used to predict the displacement field over the whole gland given sparse surface displacements. The method was validated using 3D transrectal ultrasound images of the prostates of five patients, acquired before and after expanding the balloon covering the ultrasound probe. The mean target registration error, calculated for anatomical landmarks within the gland, was 1.9mm.

Keywords: Biomechanical modelling, finite element analysis, image-guided interventions, prostate cancer, ultrasound, statistical shape modelling.

1 Introduction

Transrectal ultrasound (TRUS) imaging is used routinely to guide needle biopsy of the prostate gland as well as therapeutic interventions for prostate cancer. However, conventional (B-mode) TRUS is two-dimensional and provides very limited information on the spatial location of tumours. Consequently, performing accurate, targeted biopsy and therapy using TRUS guidance alone is practically difficult, especially for the inexperience practitioner. The ability to fuse additional information on tumour location, derived from magnetic resonance (MR) imaging or a previous biopsy, with TRUS images during a procedure therefore represents a major step towards improving the accuracy of image-guided interventions for prostate cancer, particularly as the sensitivity and specificity of functional and structural MR imaging techniques for detecting and localising prostate cancer continue to improve [1]. The fusion of detailed 3D information on the spatial location of tumours and realtime TRUS images acquired during a procedure potentially enables targeted biopsy sampling and therapies, such as radio-, brachy- and cryotherapy and, high-intensity focused US (HIFU), to be delivered at a precise location, thus avoiding important structures and minimising the risk of side effects.

F. Bello, E. Edwards (Eds.): ISBMS 2008, LNCS 5104, pp. 79–88, 2008.

Largely due to the very different intensity characteristics between MR and US images of the prostate (especially those arising from US-specific artefacts), automatic, intensity-based image registration techniques perform poorly in general. Feature-based methods are also of limited value because of the difficulty associated with extracting anatomical features from TRUS images without significant user interaction, which renders such techniques impractical in the clinical situation due to time constraints. The fact that the prostate is a deformable organ is also problematic as this means that the assumption of a rigid-body motion does not hold. Significant gland motion (including deformation) commonly takes place due to forces exerted by the bladder and rectum, patient position (e.g. supine versus lithotomy), and the placement of an US probe or MR coil into the rectum [2]. Potentially large changes in gland volume can also occur during and following interventions as a physiological response to needle insertion, ablative therapy, or both (but these sources of motion are not considered in this paper).

We have recently reported on a hybrid approach in which a patient-specific 3D model of the prostate surface (capsule), derived from an MR image, is registered automatically to 3D TRUS images [3], but the accuracy and robustness of this technique is limited by the assumption that the gland remains rigid between MR and TRUS image acquisitions. Moreover, there remains the problem of how best to estimate tissue motion within the gland given information on the motion of the gland surface. Biomechanical modelling, based on finite element analysis (FEA), has been proposed as a potential method for predicting deformation within the prostate gland [4]-[8], but, when applied directly, this method requires accurate boundary conditions, such as the displacement of the gland surface and the TRUS probe/balloon, and accurate estimates of the elastic properties of the soft tissue within the anatomical region of interest. In practice, however, considerable uncertainty exists in these parameters since typically only limited boundary condition data are available from TRUS imaging during an intervention, and the material properties are known to vary considerably between different tissue types and between patients. Chi et al., for example, report a registration error of up to 4.5mm due solely to a 30% uncertainty in material properties for a linear FE model of the prostate [4]. A patient-specific dependency on material properties is also reported for both linear and non-linear biomechanical breast tissue models [9]. Furthermore, the computation time increases rapidly with the complexity of the FE model, which can make FEA methods impractical in the surgical situation and inevitably leads to a trade-off between model accuracy and efficiency. Although dramatic reductions in computation time can be achieved using, for example, graphical processor units [10], standard FEA methods for computing deformations in heterogeneous soft-tissue are still too slow to be usefully implemented within an iterative optimisation scheme, such as the one described in this paper, which addresses the uncertainties in boundary conditions and material properties.

Here, we adopt an approach originally proposed by Davatzikos et al. [11] in which biomechanical modelling is used to simulate physically plausible gland deformations under different boundary conditions and with variable material properties. The resulting deformed prostate models are then used to train a statistical model in order to generate a compact and computationally efficient model of prostate motion that is sufficiently fast to predict an intra-gland deformation field during a surgical procedure. An important advantage of this approach is that the allowed deformation of the resulting motion model is sufficiently constrained to enable the model to be fitted to sparse measurements of the displaced gland surface, which can be practically obtained from a small number of TRUS images, acquired during an intervention.

Mohamed et al. [5] previously applied this approach to construct a statistical deformation model of a phantom prostate gland using FE simulations with variable position and orientation of the (virtual) TRUS probe as training examples. In our work, the material properties assigned to different tissue regions are also varied in the biomechanical simulations, thus relaxing the requirement for accurate knowledge of these properties and enabling a much greater range of physically plausible deformations to be captured. In the work of Alterovitz et al. [6], the intra-gland deformation field, material properties and external forces are all estimated using FEA and a geometric model based on 2D MR images. However, to the best of our knowledge, the present study is the first to report on the validation of a fully 3D model of prostate motion based on FEA and *in vivo* image data, which takes into account uncertainties in material properties and the sparseness of boundary condition data that are typically available during TRUS-guided interventions.

2 Methods

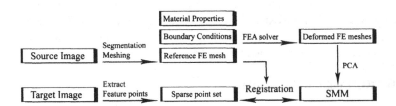

Fig. 1. Overview of the method described in this paper (see text for details). In this study, the source and target images were 3D TRUS images acquired before and after expanding the saline-filled balloon covering the TRUS probe, respectively.

2.1 Overview

A schematic overview of the method developed in this work is shown in Fig. 1. The steps involved are summarised as follows:

a) Segment the prostate gland, rectal wall, and (if visible) the pelvic bone, from the source image. (For the purposes of this study, manual segmentation was used, although, in principle, any segmentation method could be applied.)

b) From the segmented source image, build a patient-specific, 3D, reference FE mesh of the prostate gland, the rectal wall, the TRUS balloon (which ordinarily ensures acoustic coupling between the rectal wall and US transducer), and the pelvic bone.

c) Perform a large number of FEA simulations using randomly assigned material properties and boundary conditions, which correspond to different positions, orientations, and diameters of the balloon within the rectum that might be encountered during a TRUS-guided procedure.

d) Apply principal component analysis (PCA) to a set of vectors containing the simulated displacements of the node points of the gland in the FE mesh and the boundary condition parameters (which in this case define the pose and diameter of the balloon). A statistical motion model (SMM) is then formed from the principal modes of variation and constrains the deformation of the gland surface model.

e) Identify points on the surface of the gland and TRUS balloon in the target image.

f) Register the deformable gland model to the target surface points by optimising the weights of the principal modes of variation of the SMM such that the distance between the target point set and the deformed model surface is minimised. The measured diameter of the balloon in the target image is used as a further numerical constraint in this optimisation.

In order to provide an accurate gold standard for validation without implanted fiducial markers, in this study both the source and target images were 3D TRUS images. However, it is intended that in clinical practice the source image would most likely be a MR image, acquired prior to an intervention, although a CT image might also be used for radio- or brachytherapy applications. Step (a) may require significant user interaction to accurately segment the source image, whilst steps (b) and (c) are computationally intensive. Therefore, in practice, steps (a)-(d) would be performed before a procedure, whereas steps (e) and (f) would take place during the procedure (in realtime). Details of the experimental methods used to implement and validate the method outlined above are provided in the following sections.

2.2 3D TRUS Volume Acquisition

Three-dimensional TRUS images of the prostate were acquired for 5 patients undergoing a template-guided biopsy, HIFU therapy, or photodynamic therapy (PDT) for treatment of prostate cancer. All patients were recruited to clinical research studies at University College Hospital, and gave written, informed consent to participate in studies approved by the local research ethics committee. In the case of biopsy and PDT, a set of parallel transverse B-mode US images was obtained using a B-K Pro-Focus scanner (B-K Medical, Berkshire, UK) and a mechanical stepper mechanism (Tayman Medical Inc., MO, USA) to translate the probe (B-K 8658T, 5-7.5MHz transducer) axially along the rectum. Images were captured at 2mm intervals and stored on the US scanner. In the case of HIFU therapy, 3D volumes were acquired automatically using a Sonablate® 500 system (Focus Surgery, Inc., Indiana, USA). Two volumes were acquired for each patient at the start of the procedure: one with the balloon at minimal expansion, and the other after expanding the balloon by injecting saline with a syringe in order to deform the prostate gland. Expanding the balloon in this way simulates the motion of the prostate gland that might typically occur due to the presence of a TRUS probe or an endorectal MR imaging coil. The first volume was chosen as the source image for building the SMM, whilst the second (target image) was used for accuracy evaluation.

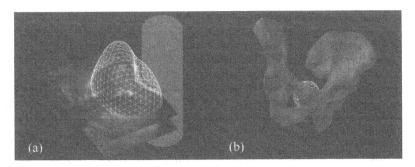

Fig. 2. (a) A triangulated surface mesh and cylinder, representing the prostate gland surface and TRUS balloon surfaces, respectively. (b) A generic model of the pelvic bone surface shown relative to the prostate gland in the FE mesh.

2.3 Finite Element Model Construction

The prostate gland was segmented in both the source and target 3D TRUS volumes by manually contouring the capsule on the acquired transverse slices. Because of known differences in compressibility between anatomical zones in the prostate, the gland was further segmented into an inner and outer gland, based on differences in echotexture on the US images [7]. A smooth surface was fitted to the surface contour points using a spherical harmonic representation from which a triangulated mesh was generated as shown in Fig. 2a. The surface of the balloon was modelled as a cylinder. The reference diameter and position of the cylinder was determined automatically by performing a least squares fit to points on the balloon surface, extracted from the original 2D TRUS images using a Canny edge detection filter. The tissue surrounding the prostate was modelled as a homogeneous block with dimensions $20{\times}20{\times}20$ cm^3.

The surface meshes and the block structure were imported into the commercial FEA software ANSYS (ANSYS Europe Ltd., Oxfordshire, UK). A FE mesh was then constructed and automatically meshed into ten-noded tetrahedral elements using trimmed parametric surfaces and Delaunay tessellation techniques provided by the software. Ten-noded tetrahedral elements were chosen because of their flexibility to generate unstructured meshes, which enables non-linear geometries to be simulated and a quadratic shape function to be employed to improve accuracy. Models for each patient consisted of between 30,000 to 40,000 elements. No badly shaped elements were generated during the model construction process. Regions corresponding to each part of the prostate gland, the rectal wall, and surrounding tissue were labelled separately and assigned different material properties (see below). All tissues were assumed to behave as isotropic, linear elastic materials. Deformations were computed using the ANSYS implementation of a pre-conditioned conjugate gradient (PCG) iterative equation solver with automatic time stepping turned on.

2.4 Boundary Conditions and Material Properties

For each case, 500 FEA simulations of prostate motion were performed using ANSYS with randomly assigned material properties and boundary conditions corresponding to different balloon (cylinder) diameters and poses. The ranges of the material properties

Table 1. Summary of the boundary conditions and material properties used for the FEA simulations. *R_0 denotes the radius of the balloon measured from the source image. **The subscripts 1-4 correspond to the inner prostate gland (1), the outer prostate gland (2), the rectal wall (3), and the surrounding tissue (4), respectively. ***The rectal wall in contact with the balloon is assumed to be nearly incompressible – i.e., $v3 = 0.49$.

Description	Parameter(s)	Range	Reference Value(s)	DOF
Balloon radius	R	$[0.9R_0, 1.5R_0]$*	R_0	1
Balloon translation	T_x, T_y, T_z	$[-5, 5]$ mm	$T_x = T_y = T_z = 0$mm	3
Balloon rotation	$\theta_x, \theta_y, \theta_z$	$[-10, 10]$ °	$\theta_x = \theta_y = \theta_z = 0°$	3
Young's modulus	E_1, E_2, E_3, E_4**	$[10, 200]$ kPa	–	4
Poisson's ratio	v_1, v_2, v_4**	$[0.30, 0.49]$	–	3***

used, given in Table 1, correspond to ranges of values reported in literature (e.g. [4]). Since the assumption of incompressibility (Poisson's ratio, $v = 0.5$) is arguably not appropriate for the prostate gland because of the likelihood of fluid loss and collapse of the urethra under compression (which explain observed changes in volume) values for both the Young's modulus and the Poisson's ratio were assigned randomly for each simulation. The expansion of the balloon was modelled by applying a radial displacement to the cylinder surface nodes. The precise position and orientation of the TRUS probe in the rectum is unknown prior to a procedure and therefore the cylinder representing the balloon surface was repositioned in each simulation by applying a random rigid-body transformation. The values of all random variables were drawn from a uniform probability distribution with upper and lower limits set by the ranges given in Table 1. In the case of the Young's modulus, only the relative values of this property are relevant.

Because only a very small part of the pelvic bone is visualised in TRUS images, it was not possible to accurately estimate its shape or location relative to the prostate (although this would be possible if MR or CT was used as the source image). For this reason, a surface model of an average male pelvis, derived from CT images [12], was used to fix the FE mesh node positions at the bone surface. The position and orientation of the bone relative to the prostate was determined using the method described in [13] (see Fig. 2b).

2.5 Statistical Motion Model

For each of M (=500) simulated gland deformations, the 3D displacement of every node in the prostate gland mesh was calculated. The components of these displacements were combined with the balloon radius and rigid-body parameters to form a vector, $\mathbf{x}$, defined (for the ith simulation) as:

$$\mathbf{x}_i = [\delta_1^T, \delta_2^T, ..., \delta_N^T, R, T_x, T_y, T_z, \theta_x, \theta_y, \theta_z]^T, \quad 1 \leq i \leq M \quad (1)$$

where N is the number of gland mesh nodes, $\delta_j^T = [\delta_x, \delta_y, \delta_z]_j$ is the 3D displacement vector for node j $(1 \leq j \leq N)$, and the remaining parameters define the radius, position and orientation of the balloon, respectively (see Table 1). The principal modes of variation in $\mathbf{x}$ were then calculated by finding the principal eigenvectors of the covariance matrix, $\mathbf{C}$, given by:

$$C = \frac{1}{M-1} \sum_{i=1}^{M} (Sx_i - \bar{x})(Sx_i - \bar{x})^T, \quad \bar{x} = \frac{1}{M} \sum_{i=1}^{M} Sx_i \tag{2}$$

In (2), S is a diagonal scaling matrix in which the diagonal element $S_{kk} = 1/\sigma_k$, where σ_k is the standard deviation of the M values for the kth element of x ($1 \leq k \leq 3N+7$). Now, if e_i is the eigenvector corresponding to the ith largest eigenvalue of C, and c_i is a scalar weight, the new mesh node co-ordinates of a deformed prostate model, contained in vector p, can be reconstructed using:

$$p = p_0 + S^{-1} \left(\bar{x} + \sum_{i=1}^{L} c_i e_i \right), \quad 1 \leq L \leq M \tag{3}$$

where p_0 contains the gland node co-ordinates in the reference mesh (derived from the source image) and the reference balloon parameters. For any $L < M$, the RMS model reconstruction error for one training simulation can be defined as:

$$\varepsilon = \sqrt{\frac{1}{n} \left(\sum_{i=L+1}^{M} c_i S^{-1} [e_i]_{(a,b)} \right)^T \left(\sum_{i=L+1}^{M} c_i S^{-1} [e_i]_{(a,b)} \right)} \tag{4}$$

where $[e_i]_{(a,b)}$ denotes the vector formed from the elements a through to b of eigenvector e_i. If $a = 1$ and $b = n = 3N$, evaluating (4) gives the RMS error in the gland node displacements, denoted here by ε_{gland}. Similarly, we can define a reconstruction error, ε_R, for the balloon radius, R, by setting $a = b = 3N+1$ and $n = 1$. In this study, L was chosen so that the SMM covered $> 99\%$ of variance in the training data and that $\varepsilon_{gland} < 0.5$mm and $\varepsilon_R < 0.5$mm, computed over all training examples.

2.6 Model Registration

Since an accurate manual 3D segmentation of the prostate surface in the target TRUS image is impractical during a procedure, a simple, clinically feasible protocol for defining a sparse set of target points was simulated in which the segmented target gland surface was resliced in 3 sagittal and 3 transverse planes, corresponding to 6 different TRUS views. Then, 6 evenly spaced points were computed on the surface contours in each slice, yielding 36 target points in total. Initial estimates for the weights, $[c_1, c_2,...,c_L]$, were computed by solving Eq. (3) with $p = [[p_0]_{(1,3N)}, R_{target}, [p_0]_{(3N+2,3N+7)}]^T$, where R_{target} is the balloon radius measured automatically in the target TRUS image. The optimal weights were then estimated using a quasi-Newton, nonlinear least-squares algorithm provided in the MATLAB Optimization Toolbox (The Mathworks Inc., MA, USA), which minimises the distance between the target points and the deformed model gland surface with the constraint $\|[p]_{(3N+1)} - R_{target}\| < 0.5$mm. Once registered, the positions of any point inside the deformed gland can be calculated by interpolating the displaced node point positions. For the purposes of comparison with a rigid registration algorithm, the source (i.e., undeformed) surface was registered to the target points using the well-known iterative closest point (ICP) algorithm.

2.7 Accuracy Validation

A number of corresponding landmarks, including cysts, calcifications and the urethra, were identified manually in the source and target TRUS volumes. The landmarks in the source image were then propagated into the target image space using the displacement field produced by the SMM following registration. For each pair of landmarks, a target registration error (TRE) was calculated, defined as the distance between manually defined and propagated landmark in the target image space.

Table 2. Target registration errors calculated for anatomical landmarks. L is the number of principal eigenmodes used in the SMM (see Equation (3)).

Case No.	Number of Landmarks	L (Variance (%))	Mean±SD TRE (mm)		
			Start	Rigid	SMM
1 (Biopsy)	12	26 (99.89 %)	5.45 ± 1.16	3.32 ± 0.98	1.86±0.76
2 (PDT)	7	28 (99.95 %)	4.32 ± 1.39	2.59 ± 0.50	1.43±0.41
3 (HIFU)	8	23 (99.92 %)	5.03 ± 0.78	2.23 ± 1.28	2.15±1.33
4 (Biopsy)	15	24 (99.92 %)	4.47 ± 1.02	2.63 ± 0.77	1.84±1.19
5 (HIFU)	6	24 (99.92 %)	6.42 ± 1.48	3.73 ± 0.98	2.27±2.22
All	48		5.03 ± 1.29	2.87 ± 1.01	1.89±1.19

3 Results

The landmark-based TREs are given in Table 2. All registrations were completed within 10s on a PC with a 2.33GHz Intel® Core™ dual CPU processor and 3GB of RAM. The time taken to compute each SMM was between 30 and 50 hours using the same computer. Inspection of the results in Table 2 reveals that in all cases the lowest TREs were achieved by using the SMM to constrain the deformable surface registration. Although a significant proportion of gland motion (~40%) was recovered using a rigid registration scheme, the SMM-constrained deformable registration recovered over 60% of the motion on average, depending on the proportion of gland deformation.

4 Discussion

The approach described in this study combines statistical motion modelling techniques with FEA to generate patient-specific, 3D deformable models of the prostate gland that deform in a manner that is physically plausible, but sufficiently well constrained that only a small number of surface points, derived from standard TRUS views, are required for registration. Since such a model can be generated before an intervention, does not require accurate estimates of material properties or intraoperative boundary conditions, such as the exact pose of the TRUS probe, and can be rapidly registered to target surface points, this method is clinically feasible.

For this study we used a commercial FEA package (ANSYS), which supports nonlinear geometry, and a linear elastic constitutive model to calculate gland deformations. Although such a model involves simplifying assumptions on the biomechanical behaviour of soft-tissue, an SMM based on a linear elastic biomechanical model (but non-linear geometry), resulted in a reduction in TRE in all of the cases in this study compared with a simple rigid registration scheme (see Table 2). To investigate the effect of varying material properties on the predicted gland displacements, we computed, for each node in each patient model, the Euclidean distance between each displaced node position, calculated using 10 different sets of randomly assigned Young's modulus and Poisson's ratio and fixed boundary conditions ($R=1.5R_0$; $T_x=T_y=T_z=0$mm; $\theta_x = \theta_y = \theta_z = 0°$), and the mean of these 10 displaced node positions. The mean value of the maximum distance (from the 10 displacements computed for each node) was 1.82mm, whilst the maximum distance was 3.85mm. These results indicate that varying the material properties has a significant effect on the displacements predicted using the biomechanical model used in this study, given that the average magnitude of the tissue displacement due to expanding the balloon was approximately 5mm (see Table 2).

Further work is required to investigate the accuracy of generating an SMM from another imaging modality, such as MR. In this case, the diameter, position and orientation of the TRUS balloon are unknown before a procedure, but we have shown how such parameters can be estimated during the registration, given an automatically measured estimate of balloon diameter. It may also be possible to obtain estimates of the other parameters during a procedure, for example, by measuring the position and orientation of the TRUS probe using a 3D tracking device, and use these measurements to further constrain the registration optimisation. The use of 3D TRUS images to build a "preoperative" SMM provided a convenient framework for the purposes of validation, since anatomical landmarks visible in the TRUS images could be accurately tracked as the balloon was expanded. In future work, we intend to use MR images to build SMMs, which raises the problem of accurately establishing corresponding landmarks within the prostate gland. This problem could be overcome by using implanted fiducial markers as targets for evaluating the registration accuracy.

Acknowledgments. The Authors would like to thank Zeike Taylor, Christine Tanner and Tim Carter for their advice on aspects of biomechanical modelling. Dean Barratt is funded as a Royal Academy of Engineering /EPSRC Research Fellow.

References

1. Kirkham, A.P., et al.: How good is MRI at detecting and characterising cancer within the prostate? Eur. Urol. 50, 1163–1174 (2006)
2. Byrne, T.E.: A review of prostate motion with considerations for the treatment of prostate cancer. Medical Dosimetry 30, 155–161 (2005)
3. Morgan, D., et al.: Registration of preoperative MR to intraoperative ultrasound images for guiding minimally-invasive prostate interventions. In: Proc. MIUA, pp. 181–185 (2007)
4. Chi, Y., et al.: A material sensitivity study on the accuracy of deformable organ registration using linear biomechanical models. Med. Phys. 33, 421–433 (2006)
5. Mohamed, A., et al.: A combined statistical and biomechanical model for estimation of intra-operative prostate deformation. In: Dohi, T., Kikinis, R. (eds.) MICCAI 2002. LNCS, vol. 2489, pp. 452–460. Springer, Heidelberg (2002)

6. Alterovitz, R., et al.: Registration of MR prostate images with biomechanical modeling and nonlinear parameter estimation. Med. Phys. 33, 446–454 (2006)
7. Bharatha, A., et al.: Evaluation of three-dimensional finite element-based deformable registration of pre-and intraoperative prostate imaging. Med. Phys. 28, 2551–2560 (2001)
8. Crouch, J.R., et al.: Automated finite element analysis for deformable registration of prostate images. IEEE Trans. on Med. Imag. 26, 1379–1390 (2007)
9. Tanner, C., et al.: Factors influencing the accuracy of biomechanical breast models. Med. Phys. 33, 1758–1769 (2006)
10. Taylor, Z.A., et al.: Real-time nonlinear finite element analysis for surgical simulation using graphics processing units. In: Ayache, N., Ourselin, S., Maeder, A. (eds.) MICCAI 2007, Part I. LNCS, vol. 4791, pp. 701–708. Springer, Heidelberg (2007)
11. Davatzikos, C., et al.: A framework for predictive modelling of anatomical deformations. IEEE Trans. Med. Image 20, 836–843 (2001)
12. Thompson, S., et al.: Use of a CT statistical deformation model for multi-modal pelvic bone segmentation. In: Proc. SPIE Medical Imaging (2008)
13. Schallenkamp, J., et al.: Prostate position relative to pelvic bony anatomy based on intra-prostatic gold markers and electronic portal imaging. Int. J. Radiation Oncology Biol. Phys. 63, 800–811 (2005)

A Real-Time Predictive Simulation of Abdominal Organ Positions Induced by Free Breathing

Alexandre Hostettler[1,2], Stéphane A. Nicolau[1], Luc Soler[1], Yves Rémond[2], and Jacques Marescaux[1]

[1] IRCAD-Hôpital Civil, Virtual-surg, Place de l'Hôpital 1,
67091 Strasbourg Cedex, France
[2] Institut de Mécanique des Fluides et des Solides, rue Boussingault 2,
67000 Strasbourg, France
{alexandre.hostettler, stephane.nicolau, luc.soler}@ircad.fr
jacques.marescaux@ircad.fr
{yves.remond}@imfs.u-strasbg.fr

Abstract. Prediction of abdominal organ positions during free breathing is a major challenge from which several medical applications could benefit. For instance, in radiotherapy it would reduce the healthy tissue irradiation. In this paper, we present a method to predict in real-time the abdominal organs position during free breathing. This method needs an abdo-thoracic preoperative CT image, a second one limited to the diaphragm zone, and a tracking of the patient skin motion. It also needs the segmentation of the skin, the viscera volume and the diaphragm in both preoperative images. First, a physical analysis of the breathing motion shows it is possible to predict abdominal organs position from the skin position and a modeling of the diaphragm motion. Then, we present our original method to compute a deformation field that considers the abdominal and thoracic breathing influence. Finally, we show on two human data that our simulation model can predict several organs position at 50 Hz with accuracy within 2-3 mm.

Keywords: Predictive simulation, patient dependant modeling, free breathing, real-time, sliding motion, skin tracking, radiotherapy, deformable model, incompressibility.

1 Introduction

Preoperative 3D CT images are used in many medical contexts. For some of them, the breathing motion is an issue that is taken into account but not compensated. For instance, in radiotherapy the dosimetry computation is performed on a static preoperative CT image although the patient breathes freely during the irradiation. Then, the planned target volume is overestimated to irradiate totally the tumour. Consequently, healthy tissue irradiations are much more important than if the tumour position was perfectly known during the breathing.

Respiratory gating techniques [10, 14, 1] are a first attempt to reduce the breathing influence. The patient is immobilized on the intervention table and his breathing is monitored to synchronize his lung volume to the one during the preoperative CT

F. Bello, E. Edwards (Eds.): ISBMS 2008, LNCS 5104, pp. 89–97, 2008.

acquisition. These methods are however constraining: they lengthen the intervention, they are uncomfortable for the patient and the repositioning accuracy, of 1.5 mm on average, can sometimes exceed 5 mm [1] if the patient is not intubated.

Therefore, a predictive method to simulate the abdomen and thorax breathing motions would be a great improvement for the previous reported application. Internal motion during a quiet breathing being between 10 and 35 mm [3], and practitioners estimate that prediction accuracy within 4 mm would bring a significant improvement to the current protocol in radiotherapy.

Solutions to simulate organ mesh deformations and interactions are numerous [4, 5, 12, 13, 7], however they require the rheological parameters of the patient organs to provide predictive results, which is currently an unsolved issue. Furthermore, in most of case, the computation time is not compatible with a real-time simulation. Consequently, the methods that were developed for realistic simulation of surgical interventions are not adapted to intra-operative applications in the operating room.

Sarrut et al. propose to simulate an artificial 3D+t CT image during a breathing cycle from two CT images acquired at inhale and exhale breath-hold. Simulated 3D+t images are built by a vector field interpolation between both CT images. This method uses a non-rigid registration to compute the displacement vectors to interpolate [11]. Although this interesting method provides a simulated CT image, it assumes that the breathing is perfectly reproducible and cyclical. Therefore, this method is limited to patients who are under general anaesthesia and intubated. Furthermore, the abdominal organs' sliding against the peritonea or the pleura is not taken into account.

Other works consider the sliding of the lung into the rib cage; unfortunately, the modeling is limited to organs situated above the diaphragmatic boundary. Furthermore, the computation time is not compatible with the real-time applications [2, 6].

Our purpose is to compute a displacement field that take the motion of the abdominal organs (such as liver, kidneys, pancreas, spleen, etc.) into account in order to provide in real time simulated CT and mesh images during free breathing using a tracking of the skin position and a modeling of the diaphragm motion.

In previous papers, we have highlighted the link between skin variations and abdominal organs positions and presented a method that only took the cranio-caudal motion into account [8, 9]. In the new model presented in this paper, the lateral and anterior-posterior motions are integrated, due to a better analysis of the diaphragm motion. This upgraded completely modify the previous modeling.

In this paper, we firstly present a summary of the whole method that describes qualitatively each key point. Secondly, we explain in detail the major parameterization steps of the method. Finally, we provide an evaluation of our method on two human clinical data showing that our simulation can be computed at 50 Hz with prediction accuracy between 2 and 3 mm.

2   Method

2.1   Global Description

Our method needs two CT acquisitions in both inspired and expired position. One of them has to be an entire abdo thoracic acquisition, whereas the second one has to be limited to the diaphragm zone. From the first acquisition, we extract the meshes that will allow us to provide the prediction of the organ position. The second one is required to model the inhomogeneous and patient dependant motion of the diaphragm during breathing. Finally, we need a device that tracks the patient skin during his free breathing.

We consider the viscera as a single organ contained in an envelop, so called the viscera envelop (the internal organ motion between themselves is negligible with respect to the global motion). The viscera envelop slides along another one, that we called internal envelop (cf. Fig. 1&2&3).

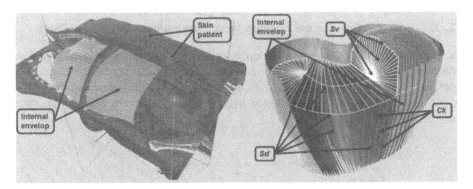

Fig. 1. The left figure represents the internal envelop and the skin. The right figure represents the internal envelop filled by the viscera volume. The viscera volume is composed of two 3D meshes, *Sv* and *Sd*, represented in two different grey levels. Cranio-caudal curves *Ck* are represented also in the right figure.

Our idea is to generate a 3D mesh *S* from the abdo-thoracic CT of the viscera volume and to compute its shape when it slides within the internal envelop. It is done using its properties of incompressibility and the modeling of the diaphragm motion. The mesh *S* is composed of two separated meshes *Sv* and *Sd* since they will be managed differently. The boundary between these two zones is an axial plan of reference situated just under the lungs, so that the tissues situated under this plan can be reasonably considered incompressible (cf. Fig. 2). We describe below how we mesh the internal envelop, *Sv* and *Sd*.

The internal envelop is composed of an axial slice stack of the same thickness. Each slice is meshed in a same radial way as described in Fig. 2. Each point of the mesh is linked to its corresponding point in the previous slice and in the next one. These broken lines of points that we call Ck are used to estimate the cranio-caudal motion (Fig. 1). During breathing, this mesh is deformed with respect to the skin position, so that to the distance between the internal envelop and the skin remains constant (see Sec. 3). Since we are only interested in the shape of the internal envelop, each mesh point is restricted to a motion in its original slice (Its Z position remains constant).

Sv is initially meshed from the abdo thoracic CT exactly like the internal envelop. During breathing, this volume mesh is deformed during two successive steps. The first step is a rough estimation of the anterior-posterior and the cranio-caudal motion of the viscera volume: firstly, the mesh point of *Sv* moves likes the point of the internal envelop mesh. Secondly, we move the mesh point along curves *Ck* to fulfil the incompressibility assumption of each slice: each slice thickness is modified to compensate its surface variation (see Sec .4). The second step allows refining the estimation of the cranio-caudal motion, and adding lateral motion to the viscera in each slice. This final estimation is computed from the diaphragm position provided by its modeling (see Sec. 5).

Sd is initially meshed so that it ensures the continuity with the mesh *Sv* (cf. Fig. 1). During breathing, mesh points are moved so that the *Sd* volume variation fit the skin volume variation under the axial plan reference (see Sec. 5).

2.2   Estimation of the Internal Envelop Mesh Position from the Skin Position

The computation of the internal envelop boundary is done in the anterior-posterior direction (cf. Fig. 2) because the configuration of the thoracic cage that contains the great part of the internal organ changes in this direction (i.e., the lateral and the crano-caudal motion of the internal envelop is neglected on comparison with the anterior-posterior motion).

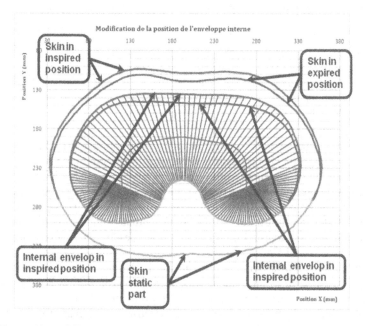

Fig. 2. The position of the skin and the internal envelop is represented in different grey levels

2.3   Rough Estimation of the Sv Mesh Motion from the Skin Position

We propose to link the cranio-caudal motion of the *Sv* mesh to the skin position. For each axial slice of *Sv*, we compute the skin volume variation below the considered slice. Since viscera are not compressible, the cranio caudal motion of the considered slice (cf. Fig. 3) compensates this variation.

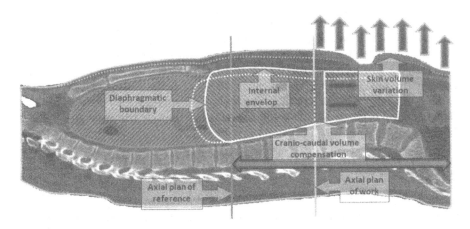

Fig. 3. The position in the inspired position is represented in continue line, whereas the expired position is represented in stippled line. The skin volume variation and the cranio-caudal average compensation are represented in colour.

2.4   Estimation of the Diaphragmatic Boundary

To assess the diaphragmatic boundary, we firstly need a modeling of the diaphragmatic motion during breathing. This modeling is obtained using both diaphragm position extracted from the two CT. Scan (cf. Fig. 4). We build in the limited CT acquisition the envelop *Sd'* that corresponds to the definition of *Sd* (the axial plan reference still the same)

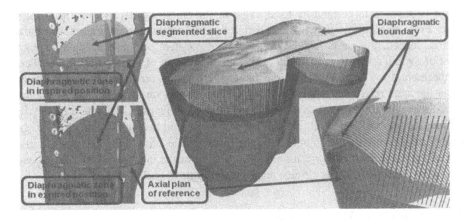

Fig. 4. Left: The segmentation of the viscera situated bellow the axial reference plan and under the diaphragmatic boundary is done in both complete initial acquisition in inspired position and the second one, limited to the diaphragmatic zone situated bellow the axial plan reference. Right: global view and detail view of the diaphragm boundary.

We mesh **Sd** and **Sd'** envelops from each sagital slice of the two CT scan and obtain two meshes, that we call **Ld** and **Ld'** (cf. Fig. 4&5). Since lateral motion is negligible, number and thickness of sagital slices in both meshes are identical, we can match each point of **Ld** to its correspondent in **Ld'**. We associate to each point in **Ld** a weight that is function of the cranio-caudal motion magnitude between the inspired and the expired position. As a result we estimate the diaphragmatic boundary from the internal envelop position and the volume variation of the skin below the axial plan of reference.

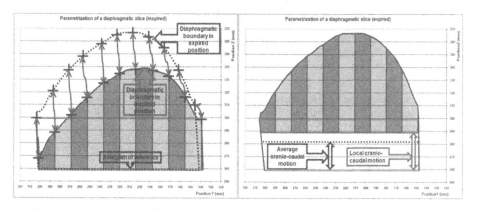

Fig. 5. Left: red double arrow represent the mesh point association in each sagital slice. The initial slice volume is represented in colour.

Therefore, since each sagital diaphragmatic slice is incompressible during breathing, it is possible to compute a local value of the cranio-caudal motion for each of them. From this information, we can refine the first estimation of the cranio-caudal motion. Indeed, the cranio-caudal motion is not constant on an axial slice for each point of **Sv**, but changes in function of the lateral position of these points in the slice. Furthermore, these local modifications have repercussions on the other slices situated under the axial reference, since the cranio-caudal motion variation around the cranio-caudal average motion is taken into account all along the meshed volume. Finally, the volume variation of the right and the left (from the middle sagital plan) part of the internal organ volume is computed and the discrepancies due to the asymmetric motion of the diaphragm is compensated by a rotation of the internal organ volume into each slide in function of the altitude and of the volume of each slice.

3 Evaluation of the Method

In order to evaluate the computation time needed by our simulation model and its prediction accuracy, we use the clinical data of two human patients (an agreement form has been signed to participate to the experiment). Thoraco-Abdominal CT acquisitions were performed in inspiration and expiration positions in order to compare the real position of the internal organs with our simulated ones. It corresponds to the classical arterial and venous phase of clinical routine thoraco-abdominal CT-scan. Then, a third acquisition, limited to diaphragmatic region is acquired in order to make our modeling of the diaphragm. The voxel size is 0.774*0.774*0.6. Our medical staff for the three CT acquisitions performed the segmentation step. In order to do not take into account the repositioning error due to the skin tracking, we make our evaluation directly using the skin position extracted from the CT acquisition. Our simulation reached a frame rate around 50 Hz on a standard laptop (core 2Duo T7700, 2.4 GHz, 2Go RAM, GeForce 8600M GT 512 MB) without implemented code optimization. A comparison of the simulated positions of the skin and internal organs with their true positions in the expired CT image is possible. Qualitative results are visible in Fig. 6&7. To evaluate quantitatively the discrepancy between each simulated and true organ, we compute for each organ the average distance between the gravity centre of each triangle of the simulated mesh and the surface of the other mesh, weighted by the triangle surface. We obtain good results, compatible with our application fields (cf. Tab 1).

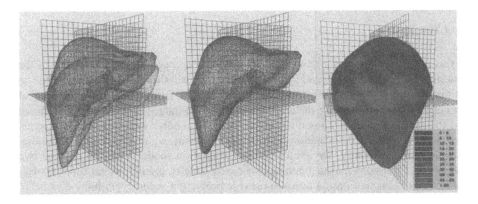

Fig. 6. Left: the liver is shown in the inspired and in the expired position. Middle: the simulated expired liver position is superimposed to the real one. Left: colours represent the distance (50=5mm) between the simulated mesh surface and the real one. The grid is made of small squares of 2 cm; therefore, we can say that the mean liver motion between the inspired position and the expired position is about 6 cm.

Table 1. Comparison between simulated meshes and real ones

Organ	Patient I	Patient II
Liver	2.4 mm	2.2 mm
Right kidney	1.9 mm	1.7 mm
Left kidney	1.7 mm	1.4 mm
Spleen	2.8 mm	2.3 mm
Pancreas		2.5 mm

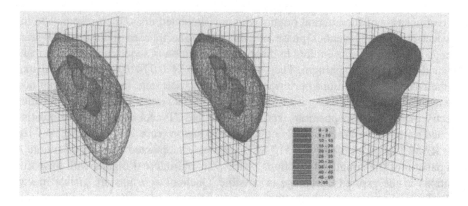

Fig. 7. Left: the right kidney is shown in the inspired and in the expired position. Middle: the simulated expired right kidney position is superimposed to the real one. Left: colours represent the distance (50=5mm) between the simulated mesh surface and the real one. The grid is made of small squares of 2 cm; therefore, we can say that the mean right kidney motion between the inspired position and the expired position is about 3 cm.

4 Conclusion

In this paper, we provide a new model to simulate and predict the movement of the abdominal organs due to breathing motion. Firstly, we showed that it is possible to estimate the displacement of main abdominal organs from a real time tracking of patient skin and a modeling of the diaphragmatic boundary that take the abdominal and the thoracic motion influence into account. Secondly, we proposed an original method to compute a deformation field from skin position and a modeling of the diaphragm motion. Finally, we evaluated our simulation model with CT data of two patients and showed that our method is not only able to predict in real-time (50 Hz) abdominal organs motions, but also to provide prediction accuracy within 2-3 mm.

References

1. Balter, J.M., Lam, K.L., McGinn, C.J., Lawrence, T.S., Ten Haken, R.K.: Improvement of CT-based treatment planning models of abdominals targets using static exhale imaging. Int. J. Radiation Oncology Biol. Phys. 41(4), 939–943 (1998)
2. Brock, K.K., Sharpe, M.B., Dawson, L.A., Kim, S.M., Jaffray, D.A.: Accuracy of finite element model-based multi-organ deformable image registration. Med. Phys. 32(6), 1647–1659 (2005)
3. Clifford, M.A., Banovac, F., Levy, E., Cleary, K.: Assessment of hepatic motion secondary to respiration for computer assisted interventions. Computer Aided Surgery 7, 291–299 (2002)
4. Cotin, S., Delingette, H., Ayache, N.: A hybrid elastic model allowing real-time cutting, deformations and force-feedback for surgery training and simulation. The Visual Computer 16(8), 437–452 (2000)

5. Delingette, H., Cotin, S., Ayache, N.: Efficient linear elastic models of soft tissues for real-time surgery simulation. In: MMVR 7 (Medicine Meets Virtual Reality), pp. 139–151 (1999)

6. Didier, A.L., Villard, P.F., Bayle, J.Y., Beuve, M., Shariat, B.: Breathing thorax simulation based on pleura behaviour and rib kinematics. Information Visualisation - MediVis, 35-40 (2007)

7. Schwartza, J.M., Denningerb, M., Rancourtb, D., Moisanc, C., Laurendeau, D.: Modelling liver tissue properties using a non-linear visco-elastic model for surgery simulation. Medical Image Analysis 9(2), 103–112 (2005)

8. Hostettler, A., Nicolau, S.A., Soler, L., Remond, Y.: Toward an accurate real time simulation of internal organ motions during free breathing from skin motion tracking and an a priori knowledge of the diaphragm motion. Int. J. of Computer assisted radiology and surgery 2(suppl. 1), 100–102 (2007)

9. Hostettler, A., Nicolau, S.A., Soler, L., Remond, Y.: Real time simulation of organ motions induced by breathing: First evaluation on patient data. In: Harders, M., Székely, G. (eds.) ISBMS 2006. LNCS, vol. 4072, pp. 9–18. Springer, Heidelberg (2006)

10. Remouchamps, V.M., Vicini, F.A., Sharpe, M.B., Kestin, L.L., Martinez, A.A., Wong, J.W.: Significant reductions in heart and lung doses using deep inspiration breathe hold with active breathing control and intensity-modulated radiation therapy for patients treated with locoregional breast irradiation. Int. J. Radiation Oncology Biol. Phys. 55, 392–406 (2003)

11. Sarrut, D., Boldea, V., Miguet, S., Ginestet, C.: Simulation of 4d ct images from deformable registration between inhale and exhale breath-hold ct scans. Medical physics 33(3), 605–617 (2006)

12. Secomb, T.W., El-Kareh, A.W.: A theoretical model for the elastic properties of very soft tissues. Biorheology 38(4), 305–317 (2001)

13. Kühnapfel, U., Çakmak, H.K., Maaß, H.: Endoscopic surgery training using virtual reality and deformable tissue simulation. Computer and Graphics 24(5), 671–682 (2000)

14. Wong, J.W., Sharpe, M.B., Jaffray, D.A., Kini, V.R., Robertson, J.M., Stromberg, J.S., Martinez, A.A.: The use of active breathing control (abc) to reduce margin for breathing motion. Int. J. Radiation Oncology Biol. Phys. 44(4), 911–919 (1999)

Effect of Friction and Material Compressibility on Deformable Modeling of Human Lung

Adil Al-Mayah, Joanne Moseley, Mike Velec, and Kristy Brock

Radiation Medicine Program, Princess Margaret Hospital, 610 University Ave. Toronto,
Ontario, M5G 2M9, Canada
adil.al-mayah@rmp.uhn.on.ca

Abstract. A three dimensional finite element model has been developed to
investigate the sliding mechanics and compressibility of human lungs of seven
lung cancer patients. The model consists of both lungs, tumor, and chest wall.
The interaction between lungs and chest cavities is modeled using surface-based
contact with coefficient of friction of 0, 0.1 and 0.2. Experimentally measured
hyperelastic material properties of the lungs are applied in the model with
different degrees of compressibility using Poisson's ratio (v) of 0.35, 0.4, 0.45
and 0.499. The analytical results are compared to actual measurements of the
bifurcation of the vessels and bronchi in the lungs and tissues. The least absolute
average error of 0.21(±0.04) cm is reached when frictionless contact surfaces
with hyperelastic material and Poisson's ratio of 0.35 and 0.4 are applied. The
error slightly changes in contact models as the coefficient of friction and
Poisson's ratio increases. However, Poisson's ratio has more effect in models
without contact surfaces where the average error changes from 0.33(±0.11) cm to
0.26(±0.07) cm as the Poisson's ratio increased from 0.35 to 0.499.

Keywords: Cancer, Contact, Friction, Lungs, Hyperelastic, Finite element.

1 Introduction

The development of the imaging modalities such as MRI, CT, SPECT and PET has
improved the quality of images and amount of patient-specific information available
in radiation therapy. More improvements have been attained by combining
multimodality images into one model of the patient for treatment planning purposes.
Furthermore, radiotherapy has advanced considerably by the introduction of image
guided radiotherapy technique where the patient is aligned in each treatment session
based on volumetric images obtained in the treatment position.

All of these techniques require image registration to relate the images obtained at
planning and treatment. Therefore, image registration plays a major role in the
accuracy of the treatment design and delivery. This registration is normally performed
by applying rigid image registration regardless of the non-rigid characteristic of
human tissues. This results in residual errors in the combination of multi-modality and
temporal images for treatment planning and delivery. Residual errors in the treatment
planning may result in uncertainties in the tumor definition and the planned dose
while residual errors in treatment delivery may cause uncertainties in the delivered
dose, potentially underdosing the tumor and overdosing the critical normal tissues.

F. Bello, E. Edwards (Eds.): ISBMS 2008, LNCS 5104, pp. 98–106, 2008.

To compensate for the deformability and movement of tissues, deformable image registration (DIR) has been introduced. By combining geometry, physics and approximation theory [1], DIR has the potential to increase the accuracy of the target delineation, calculation of the accumulated dose, and tissue response tracking while sparing healthy tissues [2,3]. A number of methods have been used in the DIR including fluid flow, optical flow, spline- and biomechanical- based models [3]. Using the biomechanical properties of tissues, finite element (FE) modeling has been applied in DIR [4,5,6,7]. The lung, which experiences large deformation due to respiration motion, is a challenging and important anatomical site to be analyzed using DIR.

Lungs were first modeled by Mead et al. 1970 [8] as a network of nonlinear springs. The first finite element model (FEM) of a lung was developed by West and Matthews [9] who investigated the behavior of human and dog lungs under its own weight. Another 3D FEM was developed using a half thoracic system [10] where internal organs lungs and heart were modeled by 23 solid elements. Both studies used dog lung material properties to simulate human lungs, however mechanical properties of the human lungs are different from those of dogs [11,12]. A 3D-FEM was developed to investigate the effect of Poisson's ratio (v) of lungs [13] using (v) of 0.25, 0.3 and 0.35. It was found the difference in displacements between the surface and solid body nodes increased as (v) increased.

In order for the lungs to function properly, they must slide relative to the chest wall and other surrounding organs. However, a little attention has been paid to model contact behaviour of lungs and other organs. Experimental investigation found the average coefficient of friction on the surface of lungs during sliding ranging between 0.078 at low speeds and 0.045 at high speed[14]. A higher value of friction of 0.3 was reported in some experiments. Contact surface was applied to morph a lung from inhale to exhale position [13,15]. Only a single lung was considered with linear elastic material properties. Previously, the authors of this study applied a contact surface model and material nonlinearity of lungs of a cancer patient [16]. The accuracy of the model improved significantly by using the contact surface and hyperelastic material properties. However only frictionless surfaces and Poisson's ratio of 0.43 were used.

In this study, the sliding behaviour of lungs' surfaces in contact with the chest cavities' is modeled simulating the actual physiological motion of the lungs. The effect of coefficient of friction on the lung deformation is investigated. Also, combined with nonlinear hyperelastic material properties of lungs, compressibility of lung's tissues using different Poisson's ratio is studied.

2   Methods and Materials

2.1   Model Description

A 3D-FEM including lungs, chest wall and tumor (Fig. 1) is developed based on 4DCT scanning images of seven lung cancer patients with different tumor locations (Fig. 2). The pixel dimensions of images are 0.098, 0.098, 0.25 cm in the Left-Right, Anterior-Posterior (AP) and Superior-Inferior (SI) directions, respectively. Four-node tetrahedral elements are used throughout the models. An inhouse multi-organ finite element based algorithm, MORFEUS, is used in building the model. MORFEUS integrates a finite element preprocessor (HYPERMESH version 7.0, Altair Engineering, Troy, MI), a finite element package (ABAQUS version 6.7, ABAQUS Inc, Pawtucket, RI) and a radiation therapy treatment planning system (PINNACLE v8.1, Philips Radiation Oncology Systems, Milpitas, CA)[6]. The model is imported to ABAQUS/CAE to add the contact surfaces.

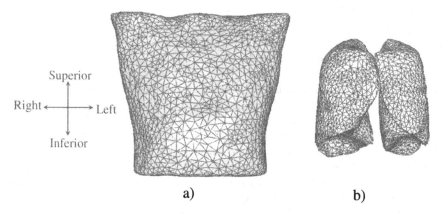

a)   b)

Fig. 1. Model components including (a) chest wall (body), and (b) lungs

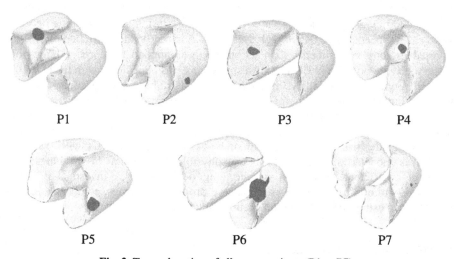

P1   P2   P3   P4

P5   P6   P7

Fig. 2. Tumor location of all seven patients (P1 to P7)

2.2 Contact Surfaces and Boundary Conditions

To allow lungs to slide against surrounding tissues, the surfaces between the left lung and left cavity, and the right lung and right cavity are modeled using surface-based contact approach. Since the pleural surface around the lungs and the pleura surrounding the chest cavities are virtually in continuous contact [17], "no separation" condition [18] is applied to the contact surfaces model to keep the surfaces in contact but not attached to each other. Three values of coefficient of friction of 0, 0.1 and 0.2 are applied on these surfaces considering the results of experimental investigation conducted by Loring et al. [16]. This parameter represents the sliding restriction of the lung inside chest cavity as a result of the viscosity of pleural fluid. Nonlinear geometry analysis is implemented to account for the expected large displacement effects. The primary and secondary states of the lungs and body are derived from scans made at the end of inhale and exhale cycles, respectively. Using HyperMorph, a guided surface projection algorithm in HyperMesh, the nodes on the inhale model are projected to the surface defined by the exhale surface of the lung. The boundary conditions are represented by the differences in location between inhale and exhale states and applied at the external of the body and surface of the chest cavities.

2.3 Material Properties

Hyperelastic material properties are applied based on the experimental test data presented by Zeng et al. [11] where ex vivo specimens were tested within 48 hours after death. Material properties, without surface tension, was the focus of the study [11]. The strain energy adopted in the experiments was as follows:

$$\rho_o W = \frac{1}{2} c \left(e^{a_1 E_x^2 + a_2 E_y^2 + 2a_4 E_x E_y} + e^{a_1 E_x^2 + a_2 E_z^2 + 2a_4 E_x E_z} + e^{a_1 E_z^2 + a_2 E_y^2 + 2a_4 E_z E_y} \right)$$

where $\rho_o W$ is the strain energy per unit volume,

c, a_1, a_2, and a_4 are constants found in the experiments with mean values of 11.8 g/cm, 0.43, 0.56 and 0.32, respectively. E_i represents strain components in a respective direction.

To simulate different compressibility of the lung tissues, Poisson's ratios of 0.35, 0.4, 0.45 and 0.499 are used. The test data are evaluated with different forms of strain energy to define the one that fits the data using the finite element package (ABAQUS/CAE). Van der Waals and Marlow are the only strain energy forms that represent the test. Marlow form is used in this study, since it provides an exact match to the test data. To compare the effect of material properties, a linear elastic material of lungs is applied with an elastic modulus of 3.74 kPa representing the linear portion of the testing data with the Poisson's ratios used with hyperelastic material properties. The body is modeled as a linearly elastic material with a modulus of elasticity of 6.0 kPa and Poisson's ratio of 0.4 [6].

2.4 Model Accuracy

The accuracy of the model deformation is evaluated by selecting landmarks from the inhale images representing the bifurcation of the vessels and bronchi as shown in Figure 3. Each of these points is tracked and located on the exhale images. Therefore, actual displacement measurements of the bifurcation points are found and later compared to the calculated displacement. The differences between the measured and calculated displacements are used to calculate the accuracy of the deformable registration.

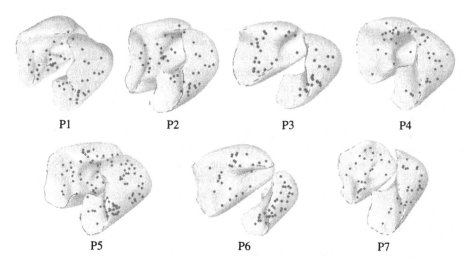

P1   P2   P3   P4

P5   P6   P7

Fig. 3. Bifurcation points distribution of all seven patients (P1 to P7)

3 Results

The effect of the variation of the coefficient of friction and Poisson's ratio are investigated by comparing the analytical results with the actual displacement of the bifurcation points. Five model configurations are used in the study, as listed in Table 1. The accuracy of the deformable registration is reported as the absolute difference (error) between the analytical results and the actual displacement. Since the largest displacement is in the Superior-Inferior (SI) direction where the displacement varies between 0.08 cm on the top of the lungs and 2.0 cm at the bottom, the largest error is expected in the SI direction as found in [16]. Therefore, the study focuses on the accuracy in the SI direction to measure the effect of mentioned parameters.

Table 1. Models description including material type, contact condition and friction

Model	Description
Elastic	Elastic material, no contact
Hyperelastic	Hyperelastic material, no contact
ContFr=0	Hyperelastic material, with contact and friction =0
ContFr=0.1	Hyperelastic material, with contact and friction =0.1
ContFr=0.2	Hyperelastic material, with contact and friction =0.2

The average absolute error of all patients (±SD) using Poisson's ratio (v) of 0.35, 0.4, 0.45 and 0.499 and coefficient of friction of 0, 0.1 and 0.2 in the SI direction of both lungs is listed in Table 2. As shown, most of the contact models with hyperelastic material properties have an average error less than 0.25 cm. However, the contact model with Poisson's ratio of 0.35 and 0.4 with a frictionless contact surface has a lower standard deviation of 0.17 ±0.03 cm (Table 3) with a success rate of 100% (Table 4) where all patients have an average error of ≤ 0.25 cm (pixel size) .

A typical effect of the coefficient of friction and Poisson's ratio is shown in Fig. 4 using the results of patient P1. The contact model with frictionless contact has the lowest average error among the rest of the models regardless of the Poisson's ratio. As the coefficient of friction increases, the error increases. This is in agreement with the average coefficient of friction values found by Loring et al. [14].

The effect of Poisson's ratio is more pronounced in models without sliding contact where the lungs are attached to the body, namely Elastic and Hyperelastic models. The average error in these two models decreases as the Poisson's ratio increases. This may be related to the fact that the incompressible material properties of the lungs makes it less compliant with respect to the displacement applied by the surrounding pleural cavity . In other words, the lungs becomes less dependent in their movement on the surrounding body. Similar finding is also reported by Villard et al. [13]. This is similar to the effect of the contact surface which allows the lungs to slide inside the chest cavity.

Table 2. Average absolute error in the SI direction of all patients

Model	Average Absolute Error (±SD among 7 patients) cm			
	$v = 0.35$	$v = 0.4$	$v = 0.45$	$v = 0.499$
Elastic	0.33(±0.11)	0.31(±0.10)	0.28(±0.09)	0.26(±0.07)
Hyperelastic	0.3(±0.09)	0.26(±0.07)	0.24(±0.06)	0.25(±0.05)
Friction = 0.0	**0.21(±0.04)**	**0.21(±0.04)**	0.22(±0. 05)	0.25(±0.05)
Friction = 0.1	0.22(±0. 05)	0.21(±0.06)	0.23(±0. 07)	0.27(±0.07)
Friction = 0.2	0.24(±0. 06)	0.23(±0. 07)	0.24(±0.07)	0.27(±0.07)

Table 3. Average standard deviation in the SI direction of all patients

Model	Average Standard Deviation (±SD among 7 patients) cm			
	$v = 0.35$	$v = 0.4$	$v = 0.45$	$v = 0.499$
Elastic	0.27 (±0.11)	0.25 (±0.11)	0.23 (±0.10)	0.20 (±0.05)
Hyperelastic	0.24 (±0. 09)	0.20 (±0.07)	0.18 (±0.06)	0.19 (±0.05)
Friction = 0.0	**0.17 (±0.03)**	**0.17 (±0.04)**	0.17 (±0.04)	0.19 (±0.05)
Friction = 0.1	0.18 (±0.05)	0.17 (±0.05)	0.18 (±0.06)	0.20 (±0.07)
Friction = 0.2	0.20 (±0.06)	0.19 (±0.06)	0.19 (±0.06)	0.21 (±0.08)

Table 4. Success rate of patients with error ≤ 0.25 cm

Model	Success rate% = (No. of patients with Error ≤ 0.25 cm / Total No. of patients) ×100			
	$v = 0.35$	$v = 0.4$	$v = 0.45$	$v = 0.499$
Elastic	14	29	43	57
Hyperelastic	29	71	57	57
Friction = 0.0	**100**	**100**	71	57
Friction = 0.1	71	71	71	43
Friction = 0.2	71	71	71	43

4 Discussion

Despite the importance of the sliding of organs for the their proper functionality, little attention has been paid to model their sliding. Besides some experimental investigation on the coefficient of friction, simulation of the coefficient of friction has not been investigated using finite element models of the lungs. In addition, the values for the compressibility of the lung reported in previous works varied substantially, however its effect on the finite element performance is mostly ignored.

The study shows that frictionless surface is the optimum condition. Therefore, the viscosity of pleural fluid and the asperties on the lungs' surface have little effect on constraining the lungs to slide freely. Poisson's ratio of 0.35 and 0.4 are the best representation of the lungs' compressibility when combined with frictionless surfaces. However, using higher Poisson's ratio without contact surface reduces the error but not as low as the optimum parameters and with lower success rate.

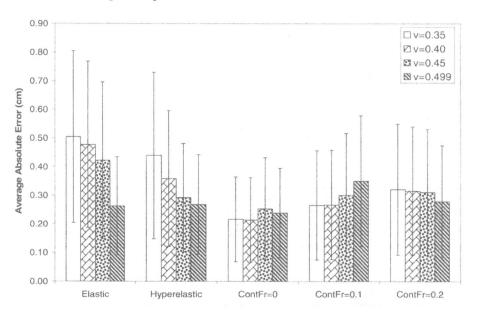

Fig. 4. Typical performance of lungs models using different Poisson's ratios (Patient P1)

5 Conclusions

The effects of the coefficient of friction and Poisson's ratio on lungs' deformation have been investigated. A 3D finite element model has been developed to simulate the left and right lungs inside the chest wall of seven cancer patients. Based on experimental test data, hyperelastic material properties of human lung tissue are applied with Poisson's ratio of 0.35, 0.4, 0.45 and 0.499. Sliding of the lungs is modeled by applying contact surfaces between the lungs and chest cavities. Three values of coefficient of friction of 0, 0.1 and 0.2 are investigated to simulate the viscosity effect of the pleural fluid. The analytical results are compared to that of actual bifurcation data in the lungs.

Generally, model accuracy increased by using sliding contact models. However, the most accurate model is the one with a frictionless sliding contact and Poisson's ratios of 0.35 and 0.4. In both models, the average error is reduced to 0.21 ± 0.04 cm. The error increases slightly as the coefficient of friction and Poisson's ratio increase. Although, the effect of Poisson's ratio is small in the contact models, it is more pronounced in the models without sliding contact surfaces where the error decreases as the Poisson's ratio increases. Work is underway to test the optimum parameters on an additional set of ten lung cancer patients.

Acknowledgements. The authors thanks Andrea Bezjak,, the Addie MacNaughton Chair in Thoracic Radiation Oncology for her assistance in obtaining the patient data. This work was supported by the National Cancer Institute of Canada - Terry Fox Foundation.

References

1. McInerney, T., Tersopoulos, D.: Deformable models in medical image analysis: a survey. Med. Image Anal., 191–108 (1996)
2. Rietzel, E., Chen, G.T.Y., Choi, N.C., Willet, C.G.: Four dimensional image-based treatment planning: target, volume segmintation and dose calculation in the presence of respiratory motion. Int. J.Radiation Oncology Biol. Phys. 61, 1535–1550 (2005)
3. Brock, K.K.: Image registration in intensity-modulated radiation therapy, image-guided radiation therapy and stereotactic body radiation therapy. Front. Radiat. Ther. Oncol. 40, 94–115 (2007)
4. Yan, D., Jaffray, D.A., Wang, J.W.: A model to accumulate fractionated dose in a deforming organ. Int. J. Radiation Oncology Biol. Phys. 44, 665–675 (1999)
5. Bharath, A., Hirose, M., Hata, N., Warfield, S.K., Ferrant, M., Zou, K.H., Suarez-Santana, E., Ruis-Alzola, J., Kikinis, R., Jolesz, F.A., Tempany, C.M.C.: Evaluation of three-dimensional finite element-based deformable registration of pre- and intraoperative prostate imaging. Med. Phys. 28, 2551–2560 (2001)
6. Brock, K.K., Sharpe, M.B., Dawson, L.A., Kim, S.M., Jaffray, D.A.: Accuracy of finite element model-based multi-organ deformable image registration. Med. Phys. 32, 1647–1659 (2005)
7. Brock, K.K., Dawson, L.A., Sharpe, M.B., Moseley, D.J., Jaffray, D.A.: Feasibility of a novel deformable image registration technique to facilitate classification, targeting, and monitoring of tumor and normal tissue. Int. J. Radiation Oncology Biol. Phys. 64, 1245–1254 (2006)
8. Mead, J., Takishima, T., Leith, D.: Stress distribution in lungs: a model of pulmonary elasticity. J. Appl. Physiol. 28, 596–608 (1970)
9. West, J.B., Matthews, F.L.: Stresses, strains, and surface pressures in the lung caused by its weight. J. Appl. Physiol. 32, 332–345 (1972)
10. Sundaram, S.H., Feng, C.C.: Finite element analysis of the human thorax. J. Biomech. 10, 505–516 (1977)
11. Zeng, Y.J., Yager, D., Fung, Y.C.: Measurement of the mechanical properties of the human lung tissue. J. Biomech. Eng. 109, 169–174 (1987)

12. De Wilde, R., Clement, J., Hellemans, J.M., Decramer, M., Demedts, M., Boving, R., Van DeWoestijne, K.P.: Model of elasticity of the human lung. J. Appl. Physiol. 51, 254–261 (1981)

13. Villard, P., Beuve, M., Shariat, B., Baudet, V., Jaillet, F.: Simulation of lung behaviour with finite elements: Influence of bio-mechanical parameters. In: Proceedings of the 3rd International Conference on Medical Information Visualisation-BioMedical Visualisation, pp. 9–14 (2005)

14. Loring, S.E., Brown, R.E., Gouldstone, A., Butler, J.P.: Lubrication regimes in mesothelial sliding. Journal of Biomechanics 38, 2390–2396 (2005)

15. Zhang, T., Orton, N.P., Rockwell Mackie, T., Paliwal, B.R.: Technical note: A novel boundary condition using contact elements for finite element based deformable image registration. Med. Phys. 31, 2412–2415 (2004)

16. Al-Mayah, A., Moseley, J., Brock, K.K.: Contact surface and material nonlinearity modeling of human lungs. Phys. Med. Biol. 53, 305–317 (2008)

17. Widmaier, E.P., Raff, H., Strang, K.T.: Vander's human physiology: the mechanisms of human body function, 10th edn. McGraw-Hill, New York (2006)

18. ABAQUS 6.7 Manual, Hibbitt, Karlsson & Sorensen, Inc.

Simulating Mechanism of Brain Injury During Closed Head Impact

Omar Halabieh[1] and Justin W.L. Wan[2]

[1] David R. Cheriton School of Computer Science, University of Waterloo,
Waterloo, ON N2L 3G1, Canada
ohalabie@uwaterloo.ca
[2] David R. Cheriton School of Computer Science, University of Waterloo,
Waterloo, ON N2L 3G1, Canada
jwlwan@uwaterloo.ca

Abstract. In this paper, we study the mechanics of the brain during closed head impact via numerical simulation. We propose a mathematical model of the human head, which consists of three layers: the rigid skull, the cerebrospinal fluid and the solid brain. The fluid behavior is governed by the Navier-Stokes equations, and the fluid and solid interact together according to the laws of mechanics. Numerical simulations are then performed on this model to simulate accident scenarios. Several theories have been proposed to explain whether the ensuing brain injury is dominantly located at the site of impact (coup injury) or at the site opposite to it (contrecoup injury). In particular, we investigate the positive pressure theory, the negative pressure theory, and the cerebrospinal fluid theory. The results of our numerical simulations together with pathological findings show that no one theory can explain the mechanics of the brain during the different types of accidents. We therefore highlight the accident scenarios under which each theory presents a consistent explanation of brain mechanics.

1 Introduction

Head injury sustained in accidents, for instance car crashes, continues to be a leading cause of death and disability. In US, millions of head injuries occurred every year. The associated socio-economic cost was estimated to be 60 billion dollars in 2000. Automobile accidents account for a large portion of these injuries, despite the use of modern protective measures such as seatbelts and airbags. For instance, data compiled by the National Center for Injury Prevention and Control in US shows that 20% of head injuries were due to motor vehicle traffic crashes. To be able to prevent head injuries and improve car safety design as well as patient diagnose and treatment, a better understanding of the mechanics of the head subjected to an impact is very important.

One particular type of brain injury of great interest is the coup-contrecoup contusion; it is frequently observed that the injury to the brain opposite to the location at which the head strikes an external object (contrecoup injury) is more severe than the injury to the brain that occurs adjacent to the impact site (coup

F. Bello, E. Edwards (Eds.): ISBMS 2008, LNCS 5104, pp. 107–118, 2008.

injury). This is known as the coup-contrecoup phenomenon. This observation has been extensively documented in pathological findings, following autopsies performed on victims [1,2,3,4,5]. Many theories have been developed to explain this counterintuitive phenomenon.

Three of the theories are: positive pressure, negative pressure and cerebrospinal fluid (CSF) theories. According to the positive pressure theory, presented by Lindenberg [1,6], as the head moves forward prior to impact, the brain tends to lag behind. Thus at the moment prior to impact, the lagging part of the brain is compressed against the skull at the contrecoup location while all the protective fluid is cushioning the coup site. At impact, a positive pressure wave amplifies the pre-existing compression of the brain tissue against the skull at the contrecoup location, causing the contrecoup injury. The negative pressure theory, presented by Russell [5,7,8], explains that once the skull is suddenly stopped at impact, the brain continues to move forward due to inertia. This results in an increase in the tensile force developing at the contrecoup site (negative pressure). This tensile force eventually leads to tearing of tissue and bridging vessels at the contrecoup location. The CSF theory was suggested by Drew [9]. It assumes that CSF is denser than brain tissue. Thus, the brain at impact is propelled to the contrecoup location due to the CSF moving forward towards the coup site.

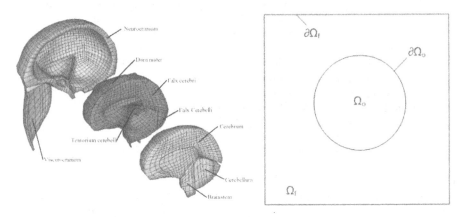

Fig. 1. (Left) A finite element model of the human head [11], (right) the proposed head model has 3 layers: skull ($\partial \Omega_f$), CSF (Ω_f), and the brain (Ω_o)

Numerical models have been developed to understand the mechanisms of this injury through simulations [11,12,13,14]. However these models have either ignored the role of the CSF or modeled it incorrectly as solid. In this paper, we propose a new head modeling approach based on the correct representation of the CSF as a fluid. More precisely, the head model is made up of three layers: the rigid skull, the CSF and finally the solid brain. Since the CSF plays such a crucial role as "a neuro-protective layer that absorbs shear energy" [7], a correct modeling of CSF is necessary to better understand the brain injury mechanism. We note that while the deformations of the skull and the brain are also important aspects of head injury, the primary focus of this paper is on how CSF affects the the motion of the brain that leads to the coup-contrecoup phenomenon. We refer the interested readers to [15,16] for mechanical response of the skull modeled as deformable shells during direct impact.

In Section 2, we present our mathematical model of the head and discuss the governing equations for the motion of each component and their interaction. Then, in Section 3, we discuss the numerical solution of the model equations. In particular, we present techniques to accurately locate the fluid-solid interface as well as numerical methods to deal with quantities near the interface. In Section 4, results of the simulations are then presented, discussed and tied in with existing theories. Finally, concluding remarks are made in Section 5.

2   Mathematical Model

There are several existing 2D/3D mathematical head models, e.g. [11,12,13,14]. Generally speaking, these models are usually based on exact geometry of the human head, see Fig. 1 (left). The modeled anatomical structures vary but usually include the skull, meningeal layers, and the brain. The skull is modeled as rigid [11] or linear elastic solid [12,13,14], the meningeal layers as linear elastic solid [11,12], and the brain as quasi-linear [12] or nonlinear viscoelastic solid [14]. There are also brain models based on viscoelastic fluid [17,18]. Then, simulations are often performed using commercial finite element softwares such as MADYMO [11] and ANSYS [13,14]. While these softwares allow the user to reliably solve complex problems, they are typically either specialized in solid or fluid calculations, rarely both at the same time. It partially explains why CSF is often omitted or modeled as a solid [11,14]. Here, we propose a mix fluid-solid head model which is more faithful to the constituent mechanical properties.

As discussed in Section 1, our primary interest is to study the coup-contrecoup phenomenon. Thus, our 2D head model as shown in Fig. 1 (right) is not intended to be elaborate nor contain accurate anatomical features. Rather, it will focus on the role of CSF plays in the coup-contrecoup phenomenon. The outer square represents the skull and the inner circle represents the brain. In between the two is the space of CSF. The particular shapes are not based on anatomical reasons but for the convenience of the numerical computation. Note that since the essence of the coup-contrecoup study is how CSF affects the motion of the brain relative to the head, we find that the particular shapes do not play a crucial role here. The following describes the model equations for each component.

Cerebrospinal fluid. CSF is modeled via the incompressible Navier-Stokes equations. These equations enforce the conservation of mass and conservation of momentum laws. Let $\mathbf{u}(x,t)$ be the velocity vector at position $\mathbf{x}$ and time t and p be the pressure. Then the motion of the fluid is governed by the equations:

$$\frac{\partial}{\partial t}\mathbf{u} + (\mathbf{u} \cdot \nabla)\mathbf{u} + \nabla p = \frac{1}{Re}\triangle\mathbf{u} + \mathbf{g} \qquad \mathbf{x} \in \Omega_f,$$

$$\nabla \cdot \mathbf{u} = 0 \qquad \mathbf{x} \in \Omega_f, \qquad (1)$$

where $\mathbf{g}$ is the gravity and Re is the so-called Reynolds number.

Fluid/solid interaction. An object immersed in a fluid is subject to a total force, **F**, given by the following equations:

$$\mathbf{F} = \int_S \boldsymbol{\tau} \cdot \mathbf{n} \, dA + \mathbf{G}, \tag{2}$$

where $\boldsymbol{\tau}$ is the hydrodynamic stress tensor, and **n** the outward-pointing normal vector on the surface S. The stress tensor $\boldsymbol{\tau}$ is given by [19]:

$$\boldsymbol{\tau} = -p\mathbf{I} + \mu(\nabla\mathbf{u} + \nabla\mathbf{u}^T), \tag{3}$$

where μ is the fluid viscosity. The first term in the equation refers to the force contribution due to the hydrostatic pressure, while the second term refers to the force contribution due to the viscosity/friction in the fluid. The net gravitational force, **G**, acting on the object is given by: $\mathbf{G} = V_o(\rho_o - \rho_f)\mathbf{g}$, where V_o is the volume of the object, ρ_o the object density and ρ_f the fluid density.

Interface modeling. As the object moves in the fluid, we model the interface motion using the level set method [20], which has achieved notable success in a wide area of applications [21,22]. The interface is represented implicitly by the zero level set of a function ϕ. For example, to represent a circle of radius r in 2D, the level set function can be defined as a cone: $\phi(x,y) = \sqrt{x^2 + y^2} - r^2$. For more general curves, numerical techniques are needed [21,22]. The region where $\phi < 0$ denotes the object and the region where $\phi > 0$ denotes the fluid. It helps us distinguish our computational domain into a fluid domain and an object domain. As the object moves, the values of the level set function ϕ are updated by solving the following equation:

$$\phi_t + \mathbf{u} \cdot \nabla\phi = 0. \tag{4}$$

Equations (1)-(4) form the governing equations for our 2D head model.

3   Numerical Solution

The numerical solution procedure involves several steps in each time stepping.

Step 1: computational domain. The Navier-Stokes equations are solved on a finite difference staggered grid consisting of grid cells. Since the equations are only defined in the fluid domain, a major issue is that we need to be able to identify fluid cells and the object cells using the values of $\{\phi_{i,j}\}$ at the grid points. We use a method similar to the one described in [23]. The idea is to determine the portion of the cell occupied by fluid and the portion occupied by the object. The dominant portion will then dictate the nature of the cell. We first give all cells the label C_F (fluid cells). Now, we identify nine different scenarios under which the cell label will be modified to C_B (object cells); see Fig. 2. The signs (+ and -) denote the sign of $\phi_{i,j}$ at the 4 corners. The top four cases mark the cell as C_B since the object occupies the major portion of the cell.

In the bottom four cases, we first determine the coordinates of the points (xa, xa2, etc) at which the interface intercepts the grid cell. This is done via linear interpolation of the two level set function values with opposite signs. Then we check to see if consequently the majority of the cell is occupied by the object. If so, the cell is labeled C_B. The last case (not shown in Fig. 2) is when four level set values are negative, thus indicating an object cell.

Fig. 2. The eight configurations for which a cell is labeled as an object cell

The boundary object cells are then identified, which can be classified into 8 types: B_N, B_W, B_E, B_S, B_NE, B_SE, B_NW, and B_SW. This labeling is based on which neighbors of an object cell are fluid. For example, B_SE denotes an object cell who's fluid neighbors are its southern and eastern neighbors. The other boundary object cell types are defined analogously.

Step 2: solve Navier-Stokes and boundary conditions. After the cell identification step, the Navier-Stokes equations are solved on the fluid domain using finite difference method. In solving the equations numerically, we also need to enforce that the fluid and object should have the same velocity at the interface. Consider a B_N cell (i, j) with a configuration illustrated in the Fig. 3. We assign values for $u_{i-1,j}$, $u_{i,j}$, and $v_{i,j}$ to satisfy the boundary condition. We first compute the coordinates of the points $y1$ and $y2$ at which the interface cuts the grid cell by linear interpolation. Then we derive the formulas for the boundary values as follows:

$$v_{i,j} = \left(\frac{h - (y1 + y2)/2}{2h - (y1 + y2)/2} \right) v_{i,j+1} + \left(\frac{h - 2h}{(y1 + y2)/2 - 2h} \right) v_{object},$$

$$u_{i-1,j} = \left(\frac{h/2 - y2}{3h/2 - y2} \right) u_{i-1,j+1} + \left(\frac{h/2 - 3h/2}{y2 - 3h/2} \right) u_{object},$$

$$u_{i,j} = \left(\frac{h/2 - y1}{3h/2 - y1} \right) u_{i,j+1} + \left(\frac{h/2 - 3h/2}{y1 - 3h/2} \right) u_{object},$$

where (u_{object}, v_{object}) is the velocity of the object. Other cell types and level set configurations are treated similarly (omitted here).

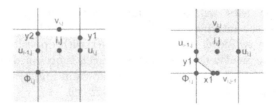

Fig. 3. B_N (left) and B_SW (right) type boundary object cells (grey: object, white: fluid)

Step 3: surface force. The force exerted by the fluid on the solid needs to be calculated. First, the equations for the stress tensor (3) are discretized on each boundary object cell. Note that the discretization depends on the type of the boundary object cell (i.e. B_N, B_W, etc). Once τ is computed, then the force exerted on this cell is: $F_{i,j} = \tau \triangle x$. The total force on the object is then computed by summing over all boundary object cells. We remark that we ignore the net gravitational force term G in (2) since physiologically the brain is tethered through the brain stem and spinal cord. Once the force F is computed, the acceleration of the object is calculated via Newton's second law: $a = F/m$, where m is the mass of the object. From the acceleration, the velocity of the object at time step $n + 1$ can be updated.

Step 4: interface motion. Finally, the object velocity computed from previous step is used to evolve the level set function according to (4) which is also solved using finite difference method.

4   Simulation and Results

The main focus of our simulations is to analyze the motion of the brain and how it depends on brain density, brain size, and head velocity. Head impact accident can be thought of as happening in two phases. The first phase is the initial acceleration of the head prior to impact (it might not be present in all accidents). The second phase is that of the actual impact. This happens when the head meets an object and comes to a full stop. The grid size for all the computations is 40 × 40. Other grid sizes have also been tested and they show similar qualitative results.

Phase 1: accelerating head prior to impact. The head is accelerating horizontally. There are two ways to incorporate this scenario in the numerical simulation. One is to set the fluid velocity u = velocity of the accelerating head on $\partial \Omega_f$. Another is to specify $u = 0$ on $\partial \Omega_f$; thus, all the computed quantities are relative to the head (not the outside viewer). In this case, the initial condition for u inside Ω_f would be negative of the moving head velocity. The former may seem natural, but the latter is what we used since it better illustrates the relative motion of the brain to the head.

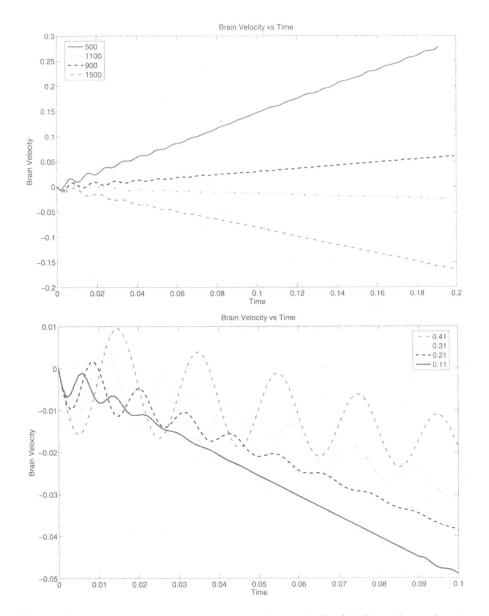

Fig. 4. Brain velocity relative to the moving head with (top) different brain density, and (bottom) different brain size

We perform the simulation to study the effect of brain density on its motion. The CSF density is fixed at $1000kg/m^3$ (the same as water), and the brain density varies from $500kg/m^3$ to $1500kg/m^3$. We remark that the actual brain density range is believed to be much smaller. This artificially wider testing range is used for the purpose of illustrating the qualitative behavior such that the differences are more visible. Fig. 4 (top) shows the simulation results. If brain tissue is denser than CSF, as the head accelerates forward, the brain lags behind. On the other hand, if CSF is denser, then the brain also moves forward. Larger density difference results in larger velocity of the brain; however, the direction of motion is unchanged.

In the second simulation, we vary the brain radius from $r = 0.11$ to 0.41. Again, the range is for qualitative illustration only and they are by no means the actual sizes of human brains. Fig. 4 (bottom) shows that larger the brain size (thus heavier the brain), slower the velocity it attains. The direction, however, is unaffected by the size. We use $1200 kg/m^3$ for the brain density in all cases and the brain always lags behind.

In the third simulation, we vary the acceleration of the moving head from $\mathbf{a} = 5m/s^2$ to $\mathbf{a} = 50m/s^2$. The brain velocity will be around 10 times larger at the final time. The direction of the brain is also unaffected. Thus, from the last two simulations, we can perform qualitative studies on brain motion with arbitrary head velocity and brain size.

Two aspects that are of importance: the brain movement and the CSF velocity field. Fig. 5 (top) shows snapshots of the brain position inside the head. The density of the brain is chosen to be less than CSF; we see that brain moving forward relative to the head, as the denser CSF lags behind. The velocity field of CSF is shown in Fig. 5 (bottom).

Phase 2: head at impact. To simulate this scenario, we set the boundary condition 0 on $\partial\Omega_f$. But since the head was moving at certain velocity $\mathbf{u}_{initial}$, so were the CSF and the brain. Thus the initial conditions for the velocities of the CSF and the brain are set to $\mathbf{u} = \mathbf{u}_{initial}$.

As in the previous phase, we study the effect of the brain density on its motion. As shown in Fig. 6 (top), if the brain is denser, as the head comes to a stop, the brain continues to more forward inside the head. On the other hand, if CSF is denser, then the brain moves backward. We also perform simulations with varying brain sizes and head speeds. The conclusions are similar to those in Phase 1; see Fig. 6 (bottom).

Verification of existing theories for closed head impact. As mentioned in Section 1, there are three main theories that have been developed to explain the coup-contrecoup phenomenon. The positive pressure and negative pressure theories implicitly assume that the CSF is less dense than brain tissue whereas the CSF theory assumes that CSF is denser than brain tissue by about 4%. In the human physiology literature, it is widely agreed upon that these two substances have very close densities; however, it is not clear which one is slightly denser than the other [24]. We will examine these theories based on the findings of our simulation results.

There are two accident scenarios that are of interest. The first scenario is when the head is accelerated by the impact; e.g. a head is being hit by a baseball bat. The second scenario is where the head is already in a state of acceleration at the moment of impact; e.g. the head of the driver hitting the steering wheel, or a

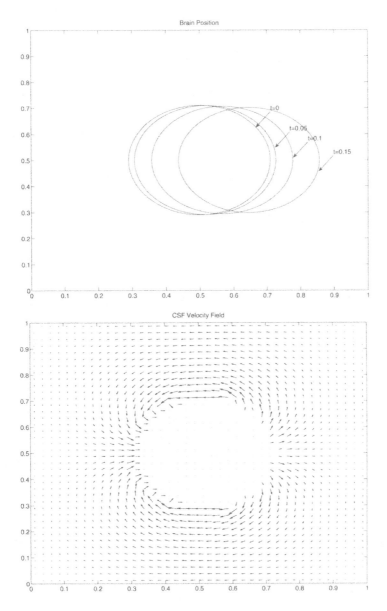

Fig. 5. (Top) Brain position at different times, (bottom) the velocity field of CSF as the brain moves to the right

fall/slip. Due to limited space, we will only discuss the second scenario. In such accidents, contrecoup injuries are usually dominant and coup injuries are minor or absent [1].

According to the positive pressure theory, as the head is accelerated prior to impact, the brain will lag behind and the less dense CSF will move to the

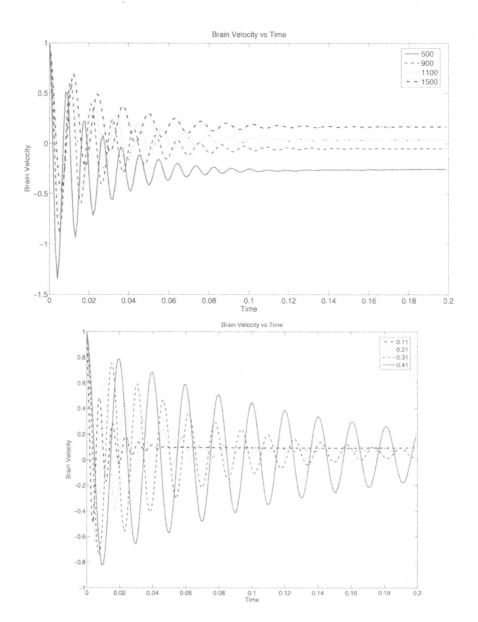

Fig. 6. Brain velocity relative to the moving head with (top) different brain density, and (bottom) different brain size

impact site. This is in fact what we observed in our simulations; see Fig. 4. Although qualitatively correct, there are drawbacks quantitatively. E.g. in a fall scenario, the acceleration that would lead to the positive pressure (lag) is about 1g, but it has been reported that the brain can withstand 40-80g's of relatively sustained acceleration without any brain injury. Thus to produce injury, the head acceleration must come from impact compression. We conclude that the initial lag in this case cannot be the main cause for the severe contrecoup injury and that the positive pressure theory is not able to explain this observation.

According to the negative pressure theory, at impact, the brain will continue to move towards the impact site. This is what we observed in our simulation if the brain tissue is denser than CSF; see Fig. 6. However, the theory does not take into account the fact that as the head is accelerating, the brain will actually have shifted towards a lagging position just prior to impact. It may counteract the forward motion right after the impact. This configuration shift prior to impact is not taken into account in the negative pressure theory.

According to the CSF theory, at impact, the brain will move towards the contrecoup site. This is what we observed in our simulation if CSF is denser than the brain; see Fig. 6. The CSF theory, however, also does not take the initial phase of head acceleration into account. In this case, the less dense brain will actually have shifted towards the impact site just prior to impact. This again might counteract the backward motion later.

5 Conclusion

We have proposed a mathematical model of the human head, which consists of the rigid skull, the CSF and the solid brain. The CSF is modeled by the Navier-Stokes equations. A numerical solution procedure is presented to describe accurate techniques to capture the fluid and solid interaction through the interface. Simulations are performed to analyze the mechanism during a head impact. The simulation results are used to verify existing theories that explain the contrecoup injury. It turns out that no one theory is able to explain all observed pathological findings in different accident scenarios. More careful studies are needed for each type of accident scenarios. These studies should further probe the pathology of head injury to identify the nature of forces causing brain injury, whether tensile or compressive or both, given the clinical observations. In addition, further investigation is required in the physiology of the human head to discern the issue of whether CSF is more or less dense than brain tissue.

References

1. Lindenberg, R., Freytag, E.: The mechanism of cerebral contusions. AMA Archives of Pathology 69, 440–469 (1960)
2. Denny-Brown, D., Russell, W.R.: Experimental cerebral concussion. Brain 64, 95–164 (1941)
3. Denny-Brown, D.: Cerebral concussion. Neurological Unit, Boston City Hospital, and the Department of Neurology, Harvard Medical School
4. Gurdjian, E., Gurdjian, E.S.: Cerebral contusions: Re-evaluation of the mechanism of their development. The Journal of Trauma 16, 35–51 (1976)
5. Russell, W.: Cerebral involvement in head injury. Brain 55, 549 (1932)
6. Lindenberg, R.: Trauma of meninges and brain. In: Minckler, J. (ed.) Pathology of the nervous system. McGraw-Hill, New York (1971)

7. Dawson, S.L., Hirsh, C.S.: The contrecoup phenomenon, reappraisal of a classical problem. Human Pathology 11, 155–166 (1980)

8. Gross, A.: A new theory on the dynamics of brain concussion and brain injury. Journal of Neurosurgery 15, 548 (1958)

9. Drew, L.B., Drew, W.E.: The contrecoup-coup phenomenon-a new understanding of the mechanism of closed head injury. Neurocritical Care 4(3), 385–390 (2004)

10. for Neuro Skills, C, http://www.neuroskills.com/edu/ceuoverview5.shtml

11. Brands, D.: Predicting brain mechanics during closed head impact. PhD thesis, Eidhoven University of Technology (2002)

12. Belingardi, G., Chiandussi, G., Gaviglio, I.: Development and validation of a new finite element model of human head. Technical report, Politecnio di Torino, Dipartimento di Meccanica, Italy (2005)

13. Chu, C.S., Lin, M.S., Huang, H.M., Lee, M.C.: Finite element analysis of cerebral contusion. Journal of Biomechanics 27(2), 187–194 (1994)

14. Chu, Y., Bottlang, M.: Finite element analysis of traumatic brain injury. In: Legacy Clinical Research and Technology Center, Portland, OR,

15. Engin, A.E.: The axisymmetric response of a fluid-filled spherical shell to a local radial impulse-a model for head injury. Journal of Biomechanics 2, 325–341 (1969)

16. Kenner, V., Goldsmith, W.: Impact on a simple physical model of the head. Journal of Biomechanics 6, 1–11 (1973)

17. Cotter, C.S., Szczyrba, P.S.I.: A viscoelastic fluid model for brain injuries. International Journal for numerical methods in fluids 40, 303–311 (2002)

18. Szczyrba, I., Burtscher, M.: On the role of ventricles in diffuse axonal injuries. In: Summer Bioengineering Conference (2003)

19. Landau, L., Lifshitz, E.: Fluid Mechanics. Pergamon Press, Oxford (1987)

20. Osher, S., Sethian, J.A.: Fronts propagating with curvature-dependent speed:algorithms based on hamilton-jacobi formulation. Journal of Computational Physics 79, 12–49 (1988)

21. Sethian, J.A.: Level Set Methods and Fast Marching Methods, 2nd edn. Cambridge University Press, Cambridge (1999)

22. Osher, S., Fedkiw, R.: Level Set Methods and Dynamic Implicit Surfaces. Springer, Heidelberg (2003)

23. Tome, M.F., McKee, S.: Gensmac: A computational marker and cell method for free surface flows in general domains. Journal of Computational Physics 110, 171–186 (1994)

24. Guyton, A., Hall, J.: Textbook of Medical Physiology, 11th edn. Elsevier Sauders, Amsterdam (2006)

Mechanism and Localization of Wall Failure During Abdominal Aortic Aneurysm Formation

Dominik Szczerba[1] , Robert McGregor[1], Krishnamurthy Muralidhar[2],
and Gábor Székely[1]

[1] Department of Electrical Engineering, ETH Zürich, Switzerland
[2] Department of Mechanical Engineering, IIT Kanpur, India

Abstract. Our previously presented model of abdominal aneurysm formation allowed to simulate aneurysm dynamics relying on a postulated initial wall failure without being able to predict the actual location of such weakening. In this study we investigate what factors can trigger pathology progression at positions eventually observed in reality. We consider mechanical effects inside the arterial wall and their possible contributions to the formation of an aneurysm. Using a computer model we demonstrate the existence of wall regions susceptible to failure due to increased oscillatory mechanical loading. We find these regions to be uniquely correlated with actually observed aneurysm locations. We demonstrate that wall fatigue and failure are probable factors influencing the formation of an abdominal aortic aneurysm.

Keywords: Abdominal Aortic Aneurysm, aneurysm formation, arterial wall mechanics, fluid-structure interaction, computational fluid dynamics.

1 Introduction

Abdominal aortic aneurysm (AAA) is a major cause of mortality in developed countries. It kills fifteen thousand people every year in the United States alone, where it is the 13th leading cause of death [1] and these numbers are likely to further increase due to ageing population. The precise causes of the disease are still not fully identified and subject to ongoing research. Its pathogenesis is complex and multifactorial, depending on many parameters such as genetic predispositions, age, lifestyle and individual aortic geometry. It occurs most frequently distal to the renal arteries and proximal to the bifurcation, suggesting that the complex flow patterns at this location may have an influence on aneurysm development.

Hemodynamic forces exerted on the arterial wall are known to induce its remodeling [2], which is usually a healthy process but can become pathological in regions with disturbed flows, such as at bifurcations. Wall shear stress (WSS) in particular and its spatial and temporal variations affect the cells of the endothelium which can lead to biochemical responses modifying wall properties, making it less able to withstand the loading imposed by the flow pressure waves. There is some debate as to whether it is low WSS, high spatial or temporal gradients of WSS or a combination of these which leads to wall lesions [3]. Much work has been done to identify areas of low or oscillating WSS in both healthy, [4] and aneurysmal [5], [6] abdominal aortic

F. Bello, E. Edwards (Eds.): ISBMS 2008, LNCS 5104, pp. 119–126, 2008.

bifurcations, but these do not consider the wall mechanics. The effect of the pressure acting on the wall is neither taken into account. Chatziprodromou et.al. [7] even quantify vascular pathology risk solely based on WSS. In some models the flow-induced wall displacement using fluid structure interaction (FSI) is also incorporated [8] [9], but this is mainly used to assess rupture risk of existing aneurysms.

FSI simulations have to rely on models, describing arterial wall mechanics. A good overview of this topic is given in [10]. The applied methods are usually isotropic simplifications of the anisotropic fibrous structure of the medial and adventitial layers, using viscoelastic constitutive laws, which in most cases are directly related to the wall's constituent parts. However, to date there are no approaches accounting for the long-term deformation of arterial walls under oscillating flow conditions.

Recently some mathematical models have been put forth to study aneurysm evolution, such as [11],investigating abdominal aortic aneurysm growth. The authors in [12] did similar work while concentrating on cerebral aneurysms. However, these and similar approaches (like [13], [14]) rely on artificial weakening at the expected location of the aneurysm. The novelty of our study is to explain the actual location of aneurysm formation and predict its growth without prior knowledge, based on purely biomechanical factors.

2 Methods

The arterial bifurcation in our computational model is represented by a simplified but biologically realistic 2D geometry in which no singularities like sharp corners are present. In order to obtain blood flow conditions, we solve the Navier-Stokes equations using a standard finite element method (FEM) discretization on an unstructured body fitted mesh. We prescribe the inflow boundary condition on the left side of the bifurcation as follows:

$$u(t) = 0.6\sin(2\pi t) + 0.2 \text{ [m/s]} \tag{1.1}$$

The exact shape of the incoming wave is more complex, but such a simplification is well justified for the purpose of our study as it captures the important *flow reversal* phase. We apply the gradient condition on the two outflows at the same time, setting the relative pressure there to zero. This results in local peak Reynolds number close to 7000. The smooth boundaries and forward bifurcation angles, however, prevent the transition of the flow into turbulent regime. Unfortunately, such boundary conditions define an ill-posed problem in the presence of complete flow reversal during a part of the heart cycle. To date, no established methods to deal with this problem exist [15]. Therefore we continue using the standard approach by simply allowing partial outflow on the inflow plane, even if we are aware that this violates the well-posedness of the problem. We minimize the side effects by making the inflow and outflow vessel branches long as compared to the size of the bifurcation apex. In addition, we relax the vertical momentum component at the inflow plane by setting its gradient to zero, instead of the more restrictive standard setting of equating its value to zero.

In reality the arterial wall is a multilayer composite material with generally nontrivial responses. Accurate modeling of its biomechanics is beyond the scope of this paper. At this point we describe the wall as an isotropic *continuum*. On the short time scale (heart cycle) the wall is assumed to displace linearly with the stresses. On the long time scale (aneurysm growth), the wall material is generalized to obey the standard viscoelastic law [16]:

$$\frac{d}{dt}\sigma(t)+\frac{E\sigma(t)}{\mu} = \frac{EE_0\epsilon(t)}{\mu}+(E+E_0)\frac{d}{dt}\epsilon(t) \tag{2.1}$$

with σ being stress, ϵ strain and μ and E the material parameters. This generalizes the material response compared to our previous model [13], where we only considered the special case of a perfectly plastic deformation.

It is frequently observed in structural materials that weak, but oscillatory forces can have dramatically more damaging effects than strong but static loads, a phenomenon known as *fatigue* [17]. It is realistic to assume that the wall (or its fibers) behave identically in this regard and that the oscillatory pressure load accumulated on the arterial wall over years might result in material fatigue. This process may in turn lead to wall failure, leading to progressive elasticity loss and ultimately to creeping, i.e., slowly progressing displacements under everlasting time averaged pressure load. We describe this phenomenon by slowly degrading parameters in the viscoelastic equation. Yield strength of such a material, i.e., stress at which material begins to deform irreversibly due to internal damage, is generally dependent on several factors, including loading frequency, load and failure history, rate of strain etc. Very generally we express this by:

$$\sigma_y = f(\nabla\times\vec{u},\frac{\partial p}{\partial t},\frac{\partial^2 l}{\partial s^2}) \tag{2.2}$$

where σ_y stands for yield stress, u and p for fluid velocity and pressure, s for the position along the bifurcation wall (see Fig. 1) and l for the wall deflection along s. The rationale for this choice is as follows. The rotations in the velocity field (first term) exert tractions on the wall and the resulting shear has already been identified as an important factor in controlling endothelial cell response, as already mentioned in the Introduction. Loading frequency (middle term) is known to influence material fatigue, and the second derivative of deflection is related to bending stress in the classical thin plate theory [17],[18]. Note that this dependence does not account for the failure history, which certainly leads to underestimating the overall effect. This is a consequence of ignoring the likely increase in failure probability resulting from the presence of already damaged fibers in the wall support network. In this work we therefore deliver *optimistic* predictions, i.e., a lower bound of the dynamics of aneurysm development. Nevertheless, Equation (2.2) stands for our basic assumption that the appearance of an aortic aneurysm is influenced by a gradual failure of the wall due to exposure to *flow induced forces* for an extended period of time, while taking into account all forces resulting in the wall due to pressure pulsation. Giving a precise form to this function will, however, not be possible without detailed experimental knowledge about the wall interior and the mechanical properties of its constituents.

To summarize our method, we start with calculating initial flow conditions in an undeformed (healthy) geometry by a computational fluid dynamics (CFD) simulation. Flow generated stresses in the arterial wall are assumed to be entirely hoop, longitudinal and bending (shear) stresses, and are obtained by computational solid mechanics (CSM). Resulting strains are calculated from a linear elastic model on the short and a viscoelastic model on the long time scale. Those strains are then used to compute wall displacements. The deformed bifurcation wall defines new boundary conditions and we periodically repeat the CFD simulation with this modified geometry. in a loop, while recording the displacements and the times needed to reach them. The required remeshing in such a fluid-structure interaction simulation is performed, once the maximal displacement exceeds 1mm.

3 Results

As seen in Figure 1, even in the healthy configuration there are clearly visible recirculations during the flow reversal phase. At the same location we find significant instabilities in pressure over the whole heart cycle. Interestingly, both phenomena occur at the expected location of an eventual real aneurysm. While the vortices are straight forward to notice, the pressure instability is more difficult to identify (Figure 1, right). Even though the global pressure difference is loading the vessel wall everywhere in the bifurcation, a special behavior can be observed at the bifurcation center. Under the assumption of linear elasticity on the short time scale, the pressure profiles in Figure 1 result almost exclusively in the (slightly) oscillating vessel radius. That is because in cylindrical coordinates hoop stress is the largest principal stress and the resulting increase in circumference is linear with growing radius. But in addition to this omnipresent gross oscillation, there is a very short bending wave along the wall at the bifurcation center, which results from a short lived longitudinal discontinuity in the pressure generated hoop stress. The strength of this bending is very small compared to the average pressure load, but - as pointed out previously - fatigue often results from small but oscillatory effects, which can only be observed at this specific location well corresponding to the usual place of the appearance of a real aneurysm. There will be likely no or little effect in a healthy individual with strong arterial fibers, but the development of an aneurysm may be triggered close to the bifurcation center, if the whole wall is weaker for some reasons. The existence of a region subject to increased

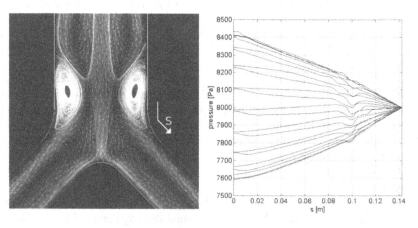

Fig. 1. Flow conditions in the unperturbed (healthy) configuration: recirculation zones close to the bifurcation apex during flow reversal (left), and time resolved pressure profiles along the bifurcation wall over one heart cycle (right)

mechanical load and the correlation of this region with the likely location of a real aneurysm clearly demonstrates the possible importance of wall mechanics in aneurysm formation, a factor which is clearly underestimated by the current literature.

4 Discussion

Based on our simulation results, two specific phenomena can be identified as major factors contributing to the actual localization of an abdominal aneurysm: vortices during flow reversal phases and longitudinal pressure instabilities along the wall. Based on currently available evidence it is hardly possible to estimate their precise contribution to aneurysm formation, but their coincidence with the usual location of the pathology is certainly remarkable. Even though it is possible that the actual triggering event is a shear induced response of the endothelium, we did not incorporate this into our model as it is already in focus of many other investigations. Instead we concentrate on mechanical aspects of the process, which received much less attention in the past. Once the weakened area is identified, we attempt to make predictions as to possible aneurysm progression entirely due to fatigue induced failure. The long term viscoelastic model parameters are currently not available to us and we had to rely on heuristic considerations in order to proceed.. If failure from fatigue is excluded and we choose large values for viscosity and elasticity, eventually no pathology of any kind can be observed at all. If we simply take the viscoelastic parameters such that a creep of half a centimeter would result in 10 years while still keeping the damage function low, we observe the wall to creep evenly all along the main axis, as a result of time and space averaged pressure load. However, if we start turning the fatigue into failure by proportionally degrading the viscoelastic parameters, we observe progression of the aneurysm as shown in Figure 2. Note that even the optimistic assumptions about the damage in (2.2) result in quite pessimistic predictions of pathology development as seen from Figure 2, right. As time passes and deformations go on, the flow and wall conditions gradually worsen (Figure 3) and subsequent deformations are clearly accelerating. If the wall has initially only slightly crept due to the fatigue induced damage, increased load immediately follows, leading to a *snowball effect*. These results already clearly demonstrate the importance of flow induced biomechanical factors in the formation and development of aneurysms.

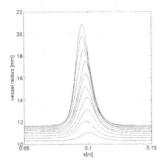 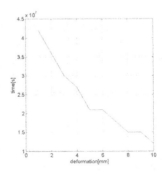

Fig. 2. Increasing wall deformation (left) and time needed to achieve every subsequent 1mm displacement (right) during aneurysm development

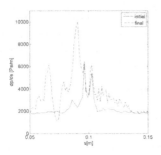 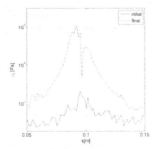

Fig. 3. Time average profiles along the wall at the initial (healthy) and final (diseased) configurations: change of pressure (left) and bending stress on a logarithmic scale (right)

5 Conclusions

On a 2D model of an unperturbed healthy abdominal aortic bifurcation we have demonstrated the existence of mechanically vulnerable regions directly corresponding to the usual anatomical position of aneurysm formation. Our findings identify fatigue, resulting wall failure and subsequent creep as a highly probable explanation for an aneurysm formation, at least in the early development stages. These results make the arbitrary postulation of wall weakening, which we have relied on in our previous work simulating the progression of the pathology [13], obsolete. The resulting simulation framework confirms our previous preliminary findings that the pathological flow and wall conditions are self sustaining. Should the vessel as a whole start losing elasticity due to some reasons like age, illness, diet or smoking, it is most likely to start creeping at actually observed locations.

Even though we are able to predict the precise initial location of an aneurysm, at this stage we can not precisely quantify its dynamics, due to the lack of detailed experimental data about the microstructure of the arterial wall and the material properties of its constituents under pathological conditions over long time periods. Histological results only show that the walls of aneurysms are often very thin and void of any structural fibers, facts in a perfect agreement with our findings [16]. The yield stress in (2.2) does not account for failure history, which is most likely an underestimation, because it does not account for the increase in failure probability caused by already damaged fibers in the network. Explicit simulations of a failing fiber network suggest a much more dramatic, exponential form. Once sufficient information about the microstructure of a damaged arterial wall are available and the precise location and orientation of the fibers can be identified, we can take advantage of this knowledge in form of an appropriate constitutive law.

First preliminary simulation on a real, patient-specific 3D geometry confirmed the qualitative validity of our 2D results. Further improvements include more advanced description of the failure process itself and the incorporation of atherosclerotic plaques into the model, which significantly influence the progression dynamics as we have already demonstrated in our previous work [13].

While this paper concentrates exclusively on the abdominal aortic bifurcation, it is possible that similar considerations are valid also for aneurysms at other anatomical locations. The exact loading patterns and the origin of fatigue will most likely vary, but the failure and subsequent creep may be universal mechanisms, significantly contributing to the pathology progression in general. The precise localization of vulnerable areas can be predicted by a patient specific simulation. This requires a dedicated processing pipeline, integrating all necessary components from image acquisition through high quality domain reconstruction and meshing to the actual computer simulation with patient specific constraints [19].

Acknowledgments. This research is supported by the Indo Swiss Joint Research Programme (ISJRP).

References

1. Gillum, R.F.: Epidemiology of aortic aneurysm in the United States. Journal of Clinical Epidemiology 48(11), 1289–1298 (1995)
2. Gibbons, G.H., Dzau, V.J.: The Emerging Concept of Vascular Remodeling. N Engl J Med. 330(20), 1431–1438 (1994)
3. Lasheras, J.C.: The Biomechanics of Arterial Aneurysms. Annual Review of Fluid Mechanics 39(1), 293–319 (2007)
4. Cheng, C.P., Herfkens, R.J., Taylor, C.A.: Abdominal aortic hemodynamic conditions in healthy subjects aged 50-70 at rest and during lower limb exercise: in vivo quantification using MRI. Atherosclerosis 168(2), 323–331 (2003)
5. Finol, E.A., Amon, C.H.: Flow-induced Wall Shear Stress in Abdominal Aortic Aneurysms: Part I - Steady Flow Hemodynamics. Computer Methods in Biomechanics and Biomedical Engineering 5(4), 309–318 (2002)
6. Raghavan, M.L., et al.: Wall stress distribution on three-dimensionally reconstructed models of human abdominal aortic aneurysm. Journal of Vascular Surgery 31(4), 760–769 (2000)
7. Chatziprodromou, I., Poulikakos, D., Ventikos, Y.: On the influence of variation in haemodynamic conditions on the generation and growth of cerebral aneurysms and atherogenesis: A computational model. Journal of Biomechanics 40(16), 3626–3640 (2007)
8. Leung, J.H., et al.: Fluid structure interaction of patient specific abdominal aortic aneurysms: a comparison with solid stress models. BioMedical Engineering OnLine 5(33) (2006)
9. Scotti, C.M., et al.: Fluid-Structure Interaction in Abdominal Aortic Aneurysms: Effects of Asymmetry and Wall Thickness. Biomed. Eng. Online 4, 64 (2005)
10. Zhao, S.Z., Xu, X.Y., Collins, M.W.: The numerical analysis of fluid-solid interactions for blood flow in arterial structures Part 1: a review of models for arterial wall behaviour. Proceedings of the Institution of Mechanical Engineers. Part H: Journal of Engineering in Medicine V212(4), 229–240 (1998)
11. Watton, P., Hill, N., Heil, M.: A mathematical model for the growth of the abdominal aortic aneurysm. Biomechanics and Modeling in Mechanobiology 3(2), 98–113 (2004)
12. Chatziprodromou, I., et al.: Haemodynamics and wall remodelling of a growing cerebral aneurysm: A computational model. Journal of Biomechanics 40(2), 412–426 (2007)
13. McGregor, R., Szczerba, D., Székely, G.: Simulation of a Healthy and a Diseased Abdominal Aorta. In: Ayache, N., Ourselin, S., Maeder, A. (eds.) MICCAI 2007, Part I. LNCS, vol. 4791, pp. 227–234. Springer, Heidelberg (2007)

14. Salsac, A.-V., Sparks, S.R., Lasheras, J.: Hemodynamic Changes Occurring during the Progressive Enlargement of Abdominal Aortic Aneurysms. Annals of Vascular Surgery 18(1), 14–21 (2004)
15. Orlanski, I.: A simple boundary condition for unbounded hyperbolic flows. Journal of Computational Physics 21(3), 251–269 (1976)
16. Humphrey, J.D., DeLange, S.: An Introduction to Biomechanics. In: Solids and Fluids, Analysis and Design. Springer, Heidelberg (2004)
17. Schijve, J.: Fatigue of Structures and Materials. Springer, Heidelberg (2008)
18. Zienkiewicz, O.C., Taylor, R.L.: The Finite Element Method
19. Szczerba, D., McGregor, R., Székely, G.: High Quality Surface Mesh Generation for Multi-physics Bio-medical Simulations. In: Computational Science – ICCS 2007, pp. 906–913 (2007)

Segmentation and 3D Reconstruction Approaches for the Design of Laparoscopic Augmented Reality Environments

P. Sánchez-González[1,2], F. Gayá[1], A.M. Cano[1,2], and E.J. Gómez[1,2]

[1] Grupo de Bioingeniería y Telemedicina (GBT), ETSIT, Universidad Politécnica de Madrid
C/ Ciudad Universitaria s/n, 28040 Madrid, Spain
{psanchez, fgaya, acano, egomez}@gbt.tfo.upm.es
http://www.gbt.tfo.upm.es
[2] Networking Research Center on Bioengineering, Biomaterials and Nanomedicine
(CIBER-BBN), Madrid, Spain

Abstract. A trend in abdominal surgery is the transition from minimally invasive surgery to surgeries where augmented reality is used. Endoscopic video images are proposed to be employed for extracting useful information to help surgeons performing the operating techniques. This work introduces an illumination model into the design of automatic segmentation algorithms and 3D reconstruction methods. Results obtained from the implementation of our methods to real images are supposed to be an initial step useful for designing new methodologies that will help surgeons operating MIS techniques.

Keywords: Laparoscopic surgery, video analysis, illumination model, segmentation, 3D reconstruction, augmented reality.

1 Introduction

Minimally Invasive Surgery (MIS) has attained a position as the best alternative of some techniques [1]. Opposed to open surgery laparoscopic techniques presents several advantages: smaller and less incisions, quick recovery, fewer post-operative complications and less pain. However, these new techniques introduce some limitations. Thus, surgeons lack depth information due to indirect vision through two-dimensional flat video displays [2][3]. A possible way to overcome this drawback is to develop new augmented reality (AR) environments to help surgeons performing surgical techniques in the operation theatre. Additional and enhanced information will be provided and fused with the real word in this environment.

The first augmented reality system applied to medicine was intended for neurosurgery [4][5]. MIS AR environments do not appear until the end of the 1990s[6]. Last proposals combine preoperative images (CT and MR are the best alternatives) and intraoperative ones (CT, MR and US). A new alternative is to use endoscopic images as intraoperative information source. This is the approach chosen in this research: the analysis of endoscopic video images for extracting useful information that can be focused on surgical scene (organs, veins…) or on surgical tools.

F. Bello, E. Edwards (Eds.): ISBMS 2008, LNCS 5104, pp. 127–134, 2008.
© Springer-Verlag Berlin Heidelberg 2008

MIS AR environments require the detachment, identification and posterior labeling of anatomical structures in the scene. Thus, it is needed an initial segmentation process of the surgical image. Endoscopic images features complicate the automation of this step. Among other features that difficult this task there are the lack of sharp borders, the transparencies and the color variability of each structure due to textures (then, a superposition of the color range exists between anatomical structures).

In addition, one of the major challenges for laparoscopic visualization is acquisition of the depth map associated with the image from the endoscopic camera. Determination of 3D scene structure from a sequence of 2D images is one of the classic problems in computer vision. Thus, lots of researchers had worked on this problem, developing numerous techniques for computing 3D structure. These techniques usually use cues from motion, stereo, shading, focus, defocus, contours, etc [8][9]. Endoscopic images features such as the absence of straight lines or planes complicate considerably reconstruction tasks. So, although there is a multitude of reconstruction approaches, it is difficult to find one that can cope with all situations. Some of their limitations are: (1) Most of existing techniques assume some physical constraints (e.g. shape from shading are used to assume the illumination and reflectance remain constant through a scene); (2) Problems in the accurate recovery of surface structure due to occlusions or camouflages (absent zones or no detected ones); (3) Computational complexity because most of the methods are iterative ones;(4) Limited application to defined textures; (5) Limited application to continuous surfaces (e.g. have to be planes).

In this research we address both challenges (3D reconstruction and automatic segmentation) based on the study of reflected lights in the scene (analyzed in the proposed illumination model) (Fig.1).

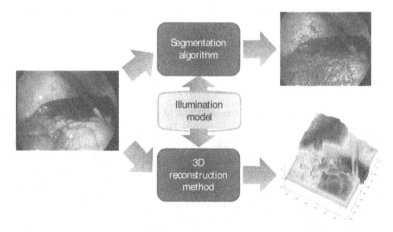

Fig. 1. Applications of the illumination model

2   Methods

This section starts with the study of the surgical scene illumination equation (2.1), and describes the two image processing algorithms designed and developed in this research: segmentation (2.2) and 3D reconstruction (2.3).

2.1 Illumination Model

In this section a study of the light reflected is made. Moreover, variables that have influence on the amount of light reflected off a 3D surface to a point of observation are analyzed.

The illumination model initial hypothesis are: (1) Laparoscopic images are characterized by the fact that light source and caption system are located at the same spacial point, the endoscope tip (see fig. 2); (2) Organs surfaces are considered as lambertian ones in surgical scenes (light is scattered equally in all directions); and (3) Indirect light intensity is considered negligible.

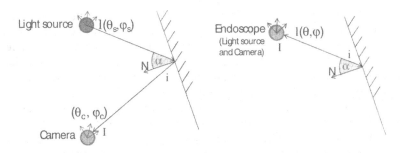

Fig. 2. General (left) and endoscope (right) scenary

Considering these hypothesis, the luminous intensity of each image pixel is computed. Let's $l(\theta_s,\varphi_s)$ be the emitted light spatial distribution in a spherical coordinates system origin located in the light source (see Fig.2). In this situation, the light emitted power L per solid angle ($[\theta_{s1},\theta_{s2}]$, $[\varphi_{s1},\varphi_{s2}]$) is calculated like:

$$L = \int_{\theta_{s1}}^{\theta_{s2}} \int_{\varphi_{s1}}^{\varphi_{s2}} l(\theta_s,\varphi_s) \cdot d\varphi_s \cdot d\theta_s \qquad (1)$$

As mentioned in hypothesis (2), an organ is treated as a lambertian surface, which intercepts the light and reflects it. The reflected light power is:

$$i = k_d \cdot l \bullet N = k_d \cdot |l| \cdot |N| \cdot \cos\alpha \qquad (2)$$

where k_d is a surface intrinsic constant and α is the angle between the light source vector (l) and the surface normal one (N) (see Fig.2).

Camera captures this reflected power. Thus, supposing the coordinates origin located in the camera focal point, power measured, per pixel in (θ_c,φ_c) direction, is:

$$I = \int_{\theta_{c1}}^{\theta_{c2}} \int_{\varphi_{c1}}^{\varphi_{c2}} i \cdot d\varphi \cdot d\theta \qquad (3)$$

Based on hypothesis (1), the two mentioned coordinates systems are fitted in the same spatial point. Therefore, equation (2) can be substituted into equation (3), which results in the simplified illumination equation of the surgical scene:

$$I = k_d \cdot L \cdot \cos\alpha \tag{4}$$

where k_d is a surface intrinsic constant, L is the emitted power per unit solid angle $([\theta_{s1},\theta_{s2}], [\varphi_{s1},\varphi_{s2}])$, and α is the angle between the light source vector and the surface normal one. This equation asserts that light intensity received in one pixel is independent of the surface-endoscope distance. This fact implies that it is not possible to determine absolute objects distance based on solely illumination information. It causes that two homothetic surfaces with similar reflectivity are undistinguished.

2.2 Segmentation Algorithm

The selected image segmentation algorithm has to produce a segmentation map. This segmentation has to be totally automatic, so methods based on manual initialization (semiautomatic methods) are not suitable for our purpose. There are many features that can be used as criteria for image segmentation. For instance, color, intensity, texture and motion are the most used. Among laparoscopic image features, intensity is considered to be the most suitable criterion in our segmentation algorithm. Due to the existence of both small and large anatomical structures, split and merge methods are well suited. However, pure merging methods are computationally expensive because they start from too small initial regions (individual pixels). So, our approach should start from a good splitting, in order to obtain the minimum number of regions as possible. Taking in mind these considerations, an automatic split and merge method is designed based on the illumination model previously explained.

A common technique for segmentation is to use gradient watershed regions, which will be used as part of the splitting stage. First a gradient magnitude image has to be built, and then watershed regions are found in this image. In this step of the process, the illumination model is used. According to the illumination equation, the image logarithm is:

$$LnI = Lnk_d + LnL + Ln\cos\alpha \tag{5}$$

Ln L term is invariant. As far as the term $Ln\cos\alpha$ is concerned, it varies in a slowly way through neighbour pixels in a same region. Similarly, k_d changes abruptly when a region ends. It makes the gradient of the logarithm be highest in region changes, which is suitable for a watershed segmentation method.

The major problem with this segmentation is that, due to the fact that k_d varies inside the organs, a certain over-segmentation is obtained. Then, we need a merging step in the segmentation process. In our merging development, Hotelling's T-square (or Mahalanobis distance) are used as the neighbour regions (or border pixels) merging metrics.

2.3 3D Reconstruction Method

The reconstruction method has to obtain a relief of the surgical scene from a monoscopic image (it is known by equation (4) that it is not possible to calculate absolute distances). Our approach is based on solely the illumination model. Then, it is questioned if the shape of an object can be reconstructed through measuring the light reflected by points form organs' surface.

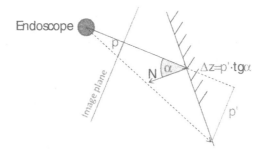

Fig. 3. Variation of the distance to the endoscope according to the tangent of the angle (α) between the visual (θ, φ) and the surface normal (N)

The problem approach is shown in fig. 3. Considering α as the angle between the light source vector (l) per solid angle (θ,φ) and the surface normal vector (N), depth variation through pixel p (Δz) is proportional to the tangent of this angle. In this way, knowing this tangent, depth variation between two neighbour pixels can be obtained.

Thus, knowing intensity slopes (tangents of α), it is possible to calculate the surfaces' shapes adding all the slopes in the gradient direction. These intensity slopes can be calculated in a two steps process:

1. Intensity gradient calculation of each pixel in a given direction of the image.
2. Division of the intensity gradient by the intensity in the direction given.

The result of this process is equal to the gradient of the image logarithm, which is the same input to watershed algorithm explained before (2.2).

Angle α is, by definition, positive. The gradient of cosine α logarithm is:

$$dLn\cos\alpha = -\frac{\sin\alpha}{\cos\alpha}\cdot d\alpha = -tg\alpha\cdot d\alpha \qquad (6)$$

The resulting sign of the equation does not originates in the *sinα* term (which is positive), but from *dα*. Therefore, it indicates the fact that α increases its value (negative sign) or decreases it (positive sign). Supposing α varies monotonically, depth variation can be accumulated, although solely depth variation absolute value is obtained.

3   Preliminary Results and Discussion

A new illumination model has been proposed for the laparoscopic images analysis. This illumination model has been used to design and develop both a segmentation algorithm and a reconstruction method.

The algorithms proposed are implemented in Matlab and computed in a Pentium Intel Core 2 Duo T7100 (2GHz RAM). Laparoscopic images were supplied by the Minimally Invasive Surgery Centre (Cáceres, Spain). Images were captured by an endoscope consisting of an Endocam 5506 camera (Richard Wolf) and an optical system Circon-ACMI 0°. Typical size of images processed is 576x768 pixels.

Sample preliminary result of laparoscopic image segmentations is shown in Fig. 4. Splitting step based on the proposed illumination model results in a start point specially adapted to laparoscopic images, distinguishing between big organs (e.g. liver) and anatomic details such as vases.

In addition, an important feature is that abrupt slope variations are not present in the surgical scene. It facilitates segmentation step due to the term cos α remains approximately constant.

Fig. 4. Segmentation preliminary results (Initial image and split image)

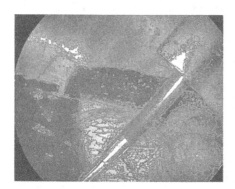

Fig. 5. (Left) Segmentation with the algorithm proposed, (Right) Watershed segmentation

Although results obtained in this study show that merging stage should be improved using other additional metrics, accuracy achieved in laparoscopic surgery images (Fig.5-Left) is higher than obtained by classic automatic segmentation methods such as typical watershed (Fig.5-Right). Moreover, with classical watershed techniques, computational cost increases considerably in merging stage (due to the great difference in number of regions after the splitting: from 3417 with our approach to 15488 with simple watershed, for image analyzed).

In addition to the previous algorithm, a 3D reconstruction method of the surgical scene has been proposed based on the illumination model. In its design, a simple light source has been considered. Nevertheless, light reflections are presented in the scene and in some situations they can be comparable to direct illumination. As red is the principal reflected color component by organs and blue and green reflected components are minimum they can be used for the developments.

Results (see Fig.6) show that although shape obtained is in a relative scale (illumination equation shows that it is not possible to extract absolute distances between organs and the endoscope through a solely use of illumination information), its main limitation is the uncertainty of the slope sign (concave/convex ambiguities).

The relative distances problem is a minor one: knowing one real depth value per continuous region and the sign of slope, it can be propagated obtaining the shape and real distance between surfaces and the endoscope.

Both problems could be addressed mixing information provided by other shape-from-X techniques such as motion (based on the matching of features from different frames) that can provide absolute distances or textures which can deal with the ambiguity of slope sign. Shape-from-shadows provides no information due to the absence of shadows, and the monoscopic nature of laparoscopic images does not allow stereo approaches.

Simplicity and low computational cost are two important characteristics of the reconstruction method proposed. Moreover, thanks to the local characteristic of this method (it uses gradient as a local operator) it is not affected by the distortions correction processing, which can be made before or after the reconstruction.

Moreover, a more complex evaluation should be done. For this evaluation a new surgical video sequence should be acquired using an optical marker, in order to get the real shape of the surface, and then compare it with our results.

 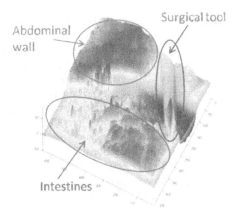

Fig.6. Original image and its depth map

4   Conclusions

Laparoscopic video analysis is a good alternative to enhance intraoperative information in order to help surgeons performing the operating techniques. In this research an illumination model has been proposed and used for designing a segmentation algorithm and a 3D reconstruction method. These are two of the main steps in the development of augmented reality environments.

In this paper it has been shown how this illumination model improves segmentation step comparing to traditional segmentation methods. Thus, accuracy and computational cost are improved.

Moreover, results obtained from application of our reconstruction method to real images show that there are some limitations that make difficult its real application. However, it can be a good starting point to design new 3D reconstruction methods that should combine illumination, texture and motion information in order to reach the final goal: to design and develop new efficient augmented reality environments, without including complex systems in the operating room.

Acknowledgments. Authors would like to thank Minimally Invasive Surgery Centre (Cáceres, Spain) for providing laparoscopic images used in this research.

References

1. Cuschieri, A.: Laparoscopic surgery: current status, issues and future developments. Surgeon 3, 125–130, 132–133, 135–138 (2005)
2. Usón, J., Pascual, S., Sanchez, F.M.: Aprendizaje y formación en cirugía laparoscópica. In: Bilbao E, Pascual S (eds.), Cirugía laparoscópica del Reflujo Gastroesofágico. Técnica de Nissen, Centro de Cirugía de Mínima Invasión, Cáceres, pp. 27-61 (2002)
3. Fuchs, K.H.: Minimally Invasive Surgery. Endoscopy 34, 154–159 (2002)
4. Kelly M.D., P.J., Kall, B., Goerss, S.: Computer-assisted stereo-taxic resection of intra-axial brain neoplasms. Journal of Neurosurgery 64, 427–439 (1986)
5. Roberts, D., Strohbehn, J., Hatch, J., Murray, W., Kettenberger, H.: A frameless stereotaxic integration of computerized tomographic imaging and the operating microscope. J. Neurosurg. 65, 545–549 (1986)
6. Freysinger, W., Gunkel, A., Thumfart, W.: Image-guided endoscopic ent surgery. European archives of Otorhinolaryngology 254, 343–346 (1997)
7. Sánchez González, P., Gayá Moreno, F., Cano González, A.M., Gómez Aguilera, E.J.: Reconstrucción tridimensional en imágenes laparoscópicas a partir de la iluminación de la escena quirúrgica. XXV Congreso Anual de la Sociedad Española de Ingeniería Biomédica (CASEIB), Cartagena, pp. 463-366 (2007)
8. Fromherz, T., Bichsel, M.: Shape from Multiple Cues: Integrating Local Brightness Information. In: Fourth International Conference for Young Computer Scientist, ICYCS 1995, Beijing, P.R. China (1995)
9. Cavanagh, P.: Reconstructing the Third Dimension: Interactions Between Color, Texture, Motion. Binocular Disparity, and Shape. CVGIP(37) 2, 171–195 (1987)

Deforming a High-Resolution Mesh in Real-Time by Mapping onto a Low-Resolution Physical Model

Hans de Visser[1], Olivier Comas[1], David Conlan[1], Sébastien Ourselin[1,2],
Josh Passenger[1], and Olivier Salvado[1]

[1] The Australian e-Health Research Centre, CSIRO, Brisbane, Australia
[2] University College London, London, United Kingdom

Abstract. For interactive surgical simulation the physical model of the soft tissue needs to be solved in real-time. This limits the attainable model density to well below the desired mesh density for visual realism. Previous work avoids this problem by using a high-resolution visual mesh mapped onto a low-resolution physical model. We apply the same approach and present an computationally cheap implementation of a known algorithm to avoid texture artefacts caused by the mapping. We also introduce a spline-based algorithm to prevent groups of high-resolution vertices, mapped to the same low-resolution triangle, from exhibiting movements in which the underlying low-resolution structure can be recognised. The resulting mapping algorithm is very efficient, mapping 54,000 vertices in 8.5 ms on the CPU and in 0.88 ms on the GPU. Consequently, the density of the high-resolution visual mesh is limited only by the detail of the CT data from which the mesh was generated.

1 Introduction

1.1 Motivation

Current rates indicate that one in twenty-two Australians will develop colorectal cancer before the age of 75[1], which is one of the highest rates in the world. Treatment and survival rates improve significantly with early detection, for which colonoscopy is regarded the optimal tool. However, colonoscopy is a difficult procedure to master with a long learning curve: it usually takes hundreds of procedures to obtain the expertise needed to perform the procedure with minimal peri- and post-operative discomfort for the patient. Patient comfort would greatly benefit from colonoscopy simulators that would allow physicians to practise the procedure to an expert level in a no-harm-to-patient environment. However, current colonoscopy training devices lack the realism in the simulation of the procedure needed to reach a level of expertise beyond the novice level [2,3,4]. The CSIRO Colonoscopy Simulation project aims to provide a more realistic environment to practise colonoscopy in, by providing a real-time photo-realistic interactive simulation with haptic feedback.

F. Bello, E. Edwards (Eds.): ISBMS 2008, LNCS 5104, pp. 135–146, 2008.

1.2 Background

The perception of realism in computer simulation depends on a number of variables. Two key factors are the way the simulation looks, and the way it behaves physically. Looks greatly benefit from a highly detailed mesh showing as much detail as possible. Accuracy of the physical behaviour of the simulation also depends on the mesh density, but also on the type of physical model chosen and on solution speed: The physical model must be solvable in "real-time", that is, at a frequency that is high enough (eg. 500 Hz) to allow haptic interaction and avoid model and contact instability. This real-time restriction in combination with available computational power sets a limit for the mesh density and model complexity achievable. Even when choosing a fairly simple physical modelling type such as mass-spring modelling, the attainable mesh density on modern PCs is lower than what would be required for visual purposes: The high-resolution colon meshes used in our simulator contain in the order of a hundred thousand nodes, whereas solving a mass-spring system in real-time on the Central Processing Unit (CPU) is only feasible for a few thousand nodes. Even when using the parallel processing power of the Graphics Processing Unit (GPU) on modern graphics cards, the number of nodes is limited to tens of thousands, not hundreds, as we also need time to detect and resolve collisions. It is noted that the mesh density required for a good perception of the physical behaviour is less than the mesh density required for a good visual perception. Also, whereas the physical model needs to be solved at a high frequency, for visualisation the mesh only needs to be updated at the rendering frequency (30Hz). Hence the logical solution to overcome the limits on mesh density would be to use a high density visual mesh driven by a lower density physical mesh.

However, for any mesh mapping to be useful, two issues need to be addressed. Firstly, irregularities in the surface distance between adjacent nodes which are attached to different parts of the physical model, must be avoided as they cause texture stretching or compression (Fig. 1). Secondly, the high-resolution nodes should not be moving as a batch in which the movement of the underlying low-resolution physical model is recognisable, or one might as well just apply the texture directly to the low-resolution model.

Mesh mapping onto physical models has been described before in soft tissue modelling and simulation, e.g. by Sun et al.[5]. Gibson and Mirtich[6] give an extensive survey of different techniques. Not all are suitable for real-time rendering however, due to their computational complexity. Mosegaard and Sørensen[7] present an elegant generic algorithm based on node normals, which with a simplification they suggest, has the potential to map very large meshes in real-time. Because of this potential, we have chosen to base our mapping on this algorithm.

To address the issue of "batch movement", a logical choice would be not to map the high-resolution mesh directly onto the physical model, but onto a smooth high-resolution surface which is derived from the physical model, for example through bicubic interpolation or mesh subdivision. Volkov and Li[8] provide an extensive overview of subdivision techniques for triangular meshes. Many more mesh interpolation schemes are found in the literature, however, we

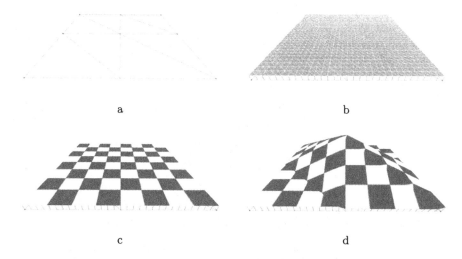

Fig. 1. Texture artefacts, such as discontinuities and stretching, can occur when a high-resolution mesh is mapped directly onto the low-resolution mesh underneath using the (low-resolution) triangle normals. *a:* low-resolution (3x3) physical model. *b:* high-resolution (31x31) mesh mapped onto the low-resolution physical model using the triangle normals. *c:* Texture applied to the high-resolution mesh. *d:* Mesh deformed by lifting the central node of the physical model. Note the texture discontinuities at the ridges where adjacent high-resolution vertices are connected to different low-resolution triangles.

could find no mention of a interpolation scheme on a per-triangle basis using only node normals. Since we already have these node normals from our mapping algorithm based on [7], and we need a computationally cheap method to meet our real-time rendering requirements, we therefore introduce a simple interpolation scheme, which uses only the node normals and is fast and accurate enough for our simulation purposes.

2   Methods

In their 2005 paper Mosegaard and Sørensen[7] present an algorithm to deal with the type of visual artefacts in high to low-resolution mapping shown in Fig. 1. The algorithm works as follows: at initialisation, each high-resolution vertex v is defined as the sum of an offset vector o and a reference point r:

$$v = o + r \ . \tag{1}$$

The reference point r is a projection of the vertex onto a triangle of the low-resolution physical model, expressed as the weighted sum of the positions of the triangle nodes p_1, p_2 and p_3:

$$r = w_1 \cdot p_1 + w_2 \cdot p_2 + w_3 \cdot p_3 \ , \tag{2}$$

$$w_1 + w_2 + w_3 = 1 \ , \tag{3}$$

in which w_1, w_2 and w_3 are the barycentric coordinates of r with regard to the triangle defined by p_1, p_2 and p_3. To ensure that the offset vector o follows the rotations and translations of the triangle it is associated with, it is expressed in the vector basis defined by the nodes of this triangle (see Fig. 2a):

$$B_t = p_3 - p_1 \ , \tag{4}$$

$$T = \text{normalise}(p_2 - p_1) \ , \tag{5}$$

$$N = \text{normalise}(T \times B_t) \ , \tag{6}$$

$$B = N \times T \ , \tag{7}$$

$$o_T = (o{\cdot}T, o{\cdot}N, o{\cdot}B) \ . \tag{8}$$

At run-time, the new position of the reference point r' is obtained by multiplying the new nodal positions p_1', p_2' and p_3' with the weighing factors w (note that throughout this paper an apostrophe ' indicates a value at run-time):

$$r' = w_1 \cdot p_1' + w_2 \cdot p_2' + w_3 \cdot p_3' \ . \tag{9}$$

The key element in preventing the visual artefact from Fig. 1 is the use of weighted node normals rather than the triangle normal when building the triangle basis at run-time (see Fig. 2b):

$$N_w' = w_1 \cdot N_1 + w_2 \cdot N_2 + w_3 \cdot N_3 \ , \tag{10}$$

$$T_t' = p_2 - p_1 \ , \tag{11}$$

$$B' = \text{normalise}(N_w' \times T_t') \ , \tag{12}$$

$$T' = \text{normalise}(B' \times N_w') \ , \tag{13}$$

$$o' = (o_{T,x} \cdot T', o_{T,y} \cdot N_w', o_{T,z} \cdot B') \ , \tag{14}$$

$$v' = o' + r' \ , \tag{15}$$

in which a node normal N_i is the normalised sum of the normals of all triangles that contain node i, see Fig. 2c.

In their paper Mosegaard and Sørensen present a general approach and do not discuss techniques for determining the reference point r and the offset vector o, apart from suggesting that the algorithm can be simplified by choosing the offset vector o along the triangle normal N. As this simplification replaces the requirement to calculate a triangle basis at run-time (11 - 14) with a simple multiplication, it provides a significant improvement in computation times and we therefore adopt this suggestion. We first calculate the length w_4 of the offset vector o by taking the dot product of the triangle normal N and the vector from (an arbitrary point on) the triangle to the vertex v:

$$w_4 = (v - p_1){\cdot}N \ . \tag{16}$$

The reference point r is then calculated easily by subtracting the offset vector o from the vertex position v:

$$o = w_4 \cdot N \ , \tag{17}$$

$$r = v - o \ . \tag{18}$$

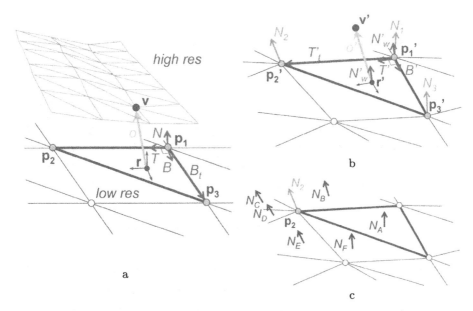

Fig. 2. *a:* Definitions of vectors and points at initialisation; *b:* Definitions of vectors and points at run-time; *c:* The node normal N_2 is defined as the normalised sum of the triangle normals $N_{A...F}$ of the 6 triangles A to F that surround node p_2

At run-time the offset vector o' is again calculated by simply multiplying the new normal N_W with the offset vector length w_4:

$$o' = w_4 \cdot N_W' \ . \tag{19}$$

However, this approach suffers from two potential inaccuracies. First of all, a vertex could potentially be mapped to the wrong triangle if there is more than one triangle for which the projection of the vertex has barycentric coordinates between 0 and 1, as shown in Fig 3a. Secondly, a slight inaccuracy occurs if the triangle normal N at initialisation is not equal to the weighted sum of node normals N_W at initialisation. In this case the vertex position calculated in the first run will deviate slightly from the initial vertex position as the projection vector changes from $o = w_4 \cdot N$ to $o' = w_4 \cdot N_W'$. Figure 3b illustrates this inaccuracy. The inaccuracies are noticeable when the physical mesh is quite coarse, and become smaller when the density of the physical mesh increases. Nevertheless, preventing them can be done easily without affecting computations at run-time, by using the following iterative process at initialisation:

1. Just like in the simplified algorithm, define r as the (perpendicular) projection of v onto the node triangle (18) by defining the offset vector o along the triangle normal N (17).
2. Express r as the weighed sum of the three nodal positions (2); in other words, calculate its barycentric coordinates w_1, w_2 and w_3.

3. Calculate the weighted sum of nodal normals N_W, just like in the run-time procedure (10).
4. Redefine the offset vector along this weighted nodes normal: $o = w_4 \cdot N_W$.
5. Calculate the new reference point r (18).
6. Repeat steps 2 - 5 until the change in the weighing factors w_1, w_2 and w_3 drops below a certain threshold (usually 0.01 or 0.001 suffices).

This iterative procedure only increases the initialisation time, but has no performance impact at run-time: for the simplified algorithm, only 9, 10, 19 and 15 need to be solved at run-time. In fact, the mapping between two meshes only needs to be calculated once and can then be stored in a file, hence the initialisation time will only be increased once.

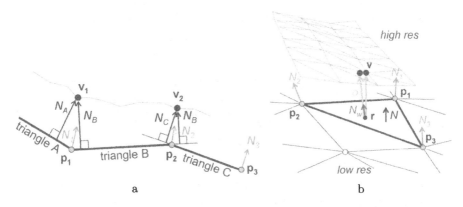

a   b

Fig. 3. *a:* As shown in this 2D example, a vertex can potentially be mapped to the wrong triangle if there is more than one triangle, or none at all, for which the perpendicular projection of the vertex has barycentric coordinates between 0 and 1, i.e. it lies within the boundaries of the triangle. For example, vertex v_1 can be projected on both triangles A and B, whereas for vertex v_2 there are no triangles for which its perpendicular projection has barycentric coordinates between 0 and 1. *b:* Inaccuracy in the position of the vertex v due to the difference between the initial triangle normal N and the initial weighted nodes normal N_W.

To address the issue of "batch movement" (Fig. 4) of the high-resolution vertices, we add a correction dw to the offset vector length w_4 based on the weighing factors w_{1-3}, which effectively means that we project v onto a curved rather than flat triangle surface (Fig. 5a). To ensure a continuous high-resolution texture, adjacent triangles must curve identically at their common edge, yet we want to calculate the curvature using solely information from the triangle itself, as the calculation of bicubic splines at run-time over the entire model, or even over just the adjacent triangles, would significantly increase computation time. This is where the weighted node normals once again prove useful, as each node normal contains information from all triangles the node is part of. Therefore we can use the node normals to ensure continuity of the curvature across the

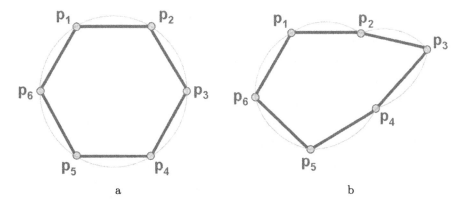

Fig. 4. *a:* Cross-section of a high-resolution tubular mesh (thin blue line) mapped onto a hexagonal physical model. *b:* Unsightly folds occur in the mesh near nodes p_2 and p_4 because all high-resolution vertices attached to the same triangle move as a batch.

triangles without the need to obtain information from adjacent triangles when calculating a triangle's curvature. By taking a weighted summation of the dot products of each node normal with the vector from that node to the projection r (Fig. 5b), a correction factor dw is obtained at very little computational cost:

$$- dw = w_1^2 \cdot N_1 \cdot (r - p_1) + w_2^2 \cdot N_2 \cdot (r - p_2) + w_3^2 \cdot N_3 \cdot (r - p_3) \ . \tag{20}$$

As detailed next, this equation can be shown to be a summation of approximate cubic Hermite splines along each of the edges of the triangle: For each triangle edge we define a standard cubic Hermite spline:

$$y(t) = (2t^3 - 3t^2 + 1) \cdot y(0) + (t^3 - 2t^2 + 1) \cdot m(0) + \dots$$
$$(-2t^3 + 3t^2) \cdot y(1) + (t^3 - t^2) \cdot m(1) \ , \tag{21}$$

where y is the height of the spline, m is the tangent (change in height) to the spline, $t = 0$ is the starting point and $t = 1$ the end point. Considering that we want the spline to go through the corner points of the triangle, the height at these points is $y(0) = y(1) = 0$ which simplifies the equation to:

$$y(t) = t \cdot (1 - t)^2 \cdot m(0) - t^2 \cdot (1 - t) \cdot m(1) \ . \tag{22}$$

Considering that for each corner point p_i the weighing factor $w_i = 1$ and the other two weighing factors are zero, since $w_1 + w_2 + w_3 = 1$, we can write this equation for each triangle edge as a function of the weighing factors. For example, as we travel along the first edge from p_1 to p_2, t increases from 0 to 1, as does w_2, while w_1 decreases from 1 to 0: $w_2 = t$ and $w_1 = 1 - t$. Therefore:

$$y_{i \to j} = w_j \cdot w_i^2 \cdot m_{i \to j}(0) - w_j^2 \cdot w_i \cdot m_{i \to j}(1) \ . \tag{23}$$

We now introduce the following approximation for the tangent m:

$$m_{i \to j}(0) \approx -N_i \cdot (p_j - p_i), \quad m_{i \to j}(1) \approx -N_j \cdot (p_j - p_i) \ . \tag{24}$$

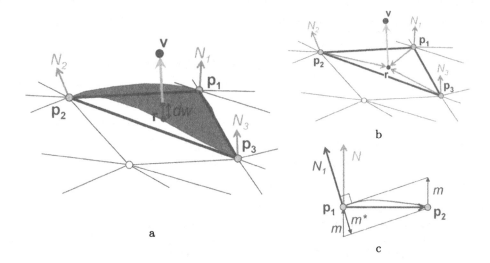

Fig. 5. *a:* Batch movement of all high-resolution vertices bound to a single low-resolution triangle is prevented by projecting the high-resolution vertices v not onto the flat triangle surface but onto a curved surface, whose curvature varies as the node normals change when the low-resolution model deforms. *b:* The correction factor dw is obtained by taking a weighted summation of the dot products of each node normal with the vector from that node to the projection r. *c:* Approximation of the tangent scalar $m \approx -m^* = -N_1 \cdot (p_2 - p_1)$.

Figure 5c shows that this is a reasonable approximation as long as the angle between the node normal N_i and the triangle normal N is small. By defining the correction factor dw as the summation of the three Hermite splines $y_{1 \rightarrow 2}$, $y_{2 \rightarrow 3}$ and $y_{3 \rightarrow 1}$, we obtain the following equation:

$$- dw = w_2 \cdot w_1^2 \cdot N_1 \cdot (p_2 - p_1) - w_2^2 \cdot w_1 \cdot N_2 \cdot (p_2 - p_1) + \dots$$
$$w_3 \cdot w_2^2 \cdot N_2 \cdot (p_3 - p_2) - w_3^2 \cdot w_2 \cdot N_3 \cdot (p_3 - p_2) + \dots$$
$$w_1 \cdot w_3^2 \cdot N_3 \cdot (p_1 - p_3) - w_1^2 \cdot w_3 \cdot N_1 \cdot (p_1 - p_3) \ , \qquad (25)$$

which can be rearranged to:

$$- dw = w_1^2 \cdot N_1 \cdot ((-w_2 - w_3) \cdot p_1 + w_2 \cdot p_2 + w_3 \cdot p_3) + \dots$$
$$w_2^2 \cdot N_2 \cdot (w_1 \cdot p_1 + (-w_1 - w_3) \cdot p_2 + w_3 \cdot p_3) + \dots$$
$$w_3^2 \cdot N_3 \cdot (w_1 \cdot p_1 + w_2 \cdot p_2 + (-w_1 - w_2) \cdot p_3) \ , \qquad (26)$$

which in combination with (2) and (3) yields the presented equation 20.

Clearly, as this correction factor dw is added to the weighing factor w_4 at runtime, w_4 need to be adjusted at initialisation by calculating and subtracting the initial correction dw. This is done after the iterative process mentioned earlier.

From the definition of the correction dw it can be recognised that the size of this correction depends on the size of the triangle, because of the correction's dependency on the vectors from the triangle corners to the reference point r. However, the (uncorrected) offset vector $o' = w_4 \cdot N'_{\mathbf{W}}$ is independent of the triangle size at run-time. In our simulation, the diameter of the tube-like colon can change considerably due to air insufflation and consequently, the triangles in the colon wall mesh can grow significantly as well. Hence, to account for these triangle size changes scaling of the weighing factor w_4 is required. There are several options for the choice of the scaling factor S. For our case both $S = \sqrt{(p_2 - p_1) \cdot (p_2 - p_1) + (p_3 - p_1) \cdot (p_3 - p_1)}$ and $S = \sqrt{|N|}$ were found to give good results, where N is the triangle normal $N = (p_2 - p_1) \times (p_3 - p_1)$ (before normalisation) which is already calculated anyway for visualisation purposes. As with the correction factor, the scaling factor needs to be calculated and applied not only at run-time: $o' = (S' \cdot w_4 + dw') \cdot N'_{\mathbf{W}}$, but also at initialisation: $w_4 = \frac{|o| - dw}{S}$.

3 Results

Figure 6a shows the effect of using the weighted node normals algorithm instead of triangle normals (Fig. 1) for calculation of the high-resolution vertex mapping. The additional effect of the curvature correction dw is depicted in Fig. 6b. The curvature correction has specific merit in the case of tubular structures with large radial deformations (Fig. 6e,f), which is particularly useful in our simulator where the colon wall can be deformed significantly by e.g. biopsy forceps. The presented algorithms have been implemented on an Intel Dual Core Xeon PC (3.0GHz, 2GB RAM) with an NVIDIA 8800GTX (768MB) graphics card. On the CPU solving a mass spring system of the colon with 1,872 masses took 10.6 ms. Mapping of a high-resolution mesh consisting of almost 54,000 vertices onto this physical model took 8.5 ms on average of which 2.0 ms were spent on the curvature correction dw. On the GPU, using NVIDIA's Compute Unified Device Architecture (CUDA)[9], the same mapping took 0.88 ms on average. Fig. 7 shows a screenshot of our simulator.

4 Discussion

The efficiency of the presented mapping approach is indicated by the fact that the mapping of almost 54,000 vertices on the CPU took less time than solving a physical model of less than 2,000 nodes. From trying out different densities for the physical model it was found that for the perception of physically accurate behaviour, the density of the physical model only needs to be as high as 10-20% of the currently used density of the high-resolution mesh. However, for visualisation the density should basically be as high as possible, which emphasises the usefulness of the mapping approach. Considering that the mapping of 54,000 vertices on the GPU takes less than a millisecond and it only needs to be done at the rendering frequency (30Hz), the achievable density of the mapped high-resolution mesh is only limited by the resolution of the input data from CT.

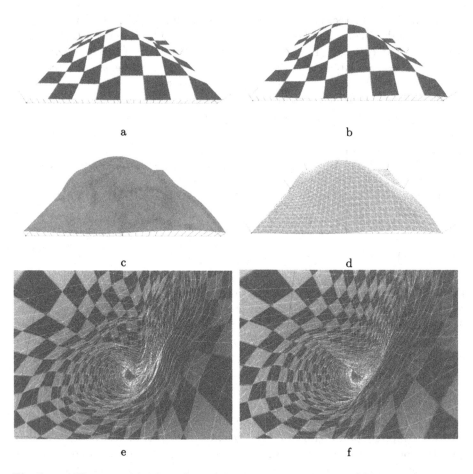

Fig. 6. *a:* The mapping algorithm of Mosegaard and Sørensen[7], used here in its simplified form as described in Eqs. 10 to 19, solves the texture artefacts seen in Fig. 1d. The node normals used in the algorithm are shown as long vertical lines. *b:* Adding the curvature correction *dw* to this mapping algorithm solves the "batch movement" of the high-resolution vertices such that the underlying low-resolution structure is no longer clearly recognisable. *c:* When applying a more gradual flesh texture rather than the checkerboard texture, the underlying structure becomes even harder to recognise. *d:* The contribution of the curvature correction *dw* shown separately. *e:* Inside a heavily deformed tube without curvature correction *dw*. Note the folds running from the centre to the top and from the centre to just below the top-left corner. *f:* The same tube with curvature correction *dw*, which makes the folds much less sharp.

We presented an iterative method to provide a consistent initial configuration for the mapping algorithm of Mosegaard and Sørensen[7]. The iterative process usually converges within 5 steps when a threshold of 0.001 is used but a safeguard has been implemented to end the process after a maximum of 20 steps. This process increases the total initialisation time for the mapping of 54,000 vertices

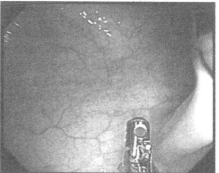

Fig. 7. Screenshot from our colonoscopy simulator (left) compared to a screenshot from a real colonoscopy (right).

from 4.5 to 6.9 sec. However, building of the map only needs to be done once and can then be saved to file, hence subsequent initialisation only takes 0.2 sec.

The presented curvature correction to prevent "batch movement" is a computationally cheap way to visually improve mesh deformation. It is particularly useful when the underlying physical model is very coarse, such as the 3x3 planar model (Fig. 6b) or a tubular model (Figs. 4, 6e and 6f) like a colon. Logically, the effect of the curvature correction diminishes when the density of the physical model approaches that of the high-resolution mesh.

When observing Fig. 5c it can be argued that instead of approximating the tangent, it can be easily calculated exactly by dividing by $N \cdot N_1$. However, this introduces an extra dot product and division, which would slow down the calculation, as well as a potential risk of "division by zero" errors against which we would need to safeguard, but more importantly it was found not to cause any noticeable improvement and was therefore left out.

It is noticed in Figs. 6b and 6d that the curvature is not always completely smooth over the edges of adjacent triangles as the presented equation 20 only ensures that the offset dw is continuous over the triangle edge, but it does not provide that same guarantee for its derivative in the direction perpendicular to the edge. A more complex correction that would have a continuous derivative over the triangle edge could be derived from the node normals, for example by using higher-order weighing of the Hermite splines instead of simple addition (Eq. 25), or from information from the adjacent triangle by using the node normal from the node opposite the edge in question. This would likely cause an increase in computation time which could jeopardise our real-time rendering requirements. Also, we deemed the potential improvement unnecessary for our purpose, as Fig. 6c already illustrates that with the type of textures we use the current correction provides a mapping in which the underlying physical model is virtually unrecognisable.

5 Conclusion

We have presented an algorithm for mapping of a high-resolution mesh onto a low-resolution physical model, which adds an iterative preprocessing step to an known algorithm [7], and a spline-based correction to improve visual appearance. Mesh mapping has proven to be a very efficient approach in our strive towards of a photo-realistic interactive colonoscopy simulator. The very low computational cost of the mapping means that currently the density of the applied mesh for visualisation is only limited by the resolution of the input data from CT.

Acknowledgements

This research was funded by the Preventative Health Flagship of the Commonwealth Scientific and Industrial Research Organisation (CSIRO) in Australia.

References

1. Australian Institute of Health & Welfare: Cancer in Australia: an overview (accessed 06-03-2008),
   http://www.aihw.gov.au/publications/index.cfm/title/10476
2. Cohen, J.: The First International Conference on Endoscopy Simulation: Consensus Statement. Gastrointestinal Endoscopy 16, 583–591 (2006)
3. Gerson, L.: Evidence-Based Assessment of Endoscopic Simulators for Training. Gastrointestinal Endoscopy 16, 489–509 (2006)
4. Sedlack, R., Kolars, J.: Computer Simulator Training Enhances the Competency of Gastroenterology Fellows at Colonoscopy: Results of a Pilot Study. American Journal of Gastroenterology 99, 33–37 (2004)
5. Sun, W., Hilton, A., Smith, A.R., Illingworth, J.: Layered Animation of Captured Data. Visual Computer 17, 457–474 (2001)
6. Gibson, S.F.F., Mirtich, B.: A survey of deformable modeling in computer graphics. Mitsubishi Electric Research Laboratory, report no. TR97-19 (1997)
7. Mosegaard, J., Sørensen, T.: Real-time Deformation of Detailed Geometry Based on Mappings to a Less Detailed Physical Simulation on the GPU. In: Proceedings of Eurographics Workshop on Virtual Environments, vol. 11, pp. 105–111 (2005)
8. Volkov, V., Li, L.: Real-Time Refinement and Simplification of Adaptive Triangular Meshes. IEEE Visualization, 155–162 (2003)
9. NVIDIA: CUDA Programming Guide 1.1. (Accessed 12-03-2008),
   http://www.nvidia.com/object/cuda_develop.html

Versatile Design of Changing Mesh Topologies for Surgery Simulation

Barbara André and Hervé Delingette

INRIA Sophia Antipolis, Asclepios Research Project,
2004 route des Lucioles BP 93, 06902 Sophia Antipolis Cedex, France
{barbara.andre,herve.delingette}@sophia.inria.fr
http://www-sop.inria.fr/asclepios

Abstract. In the context of surgery simulation, this paper presents a generic and efficient solution to handle topological changes on deformable meshes under real-time constraints implemented in the SOFA [4] platform. The proposed design is based on a simulation tree gathering software components acting on a mesh. The mesh topology is described by a topological component which also provides algorithms for performing topological changes (cutting, refinement). An important aspect of the design is that mesh related data is not centralized in the mesh data structure but stored in each dedicated component. Furthermore, topological changes are handled in a transparent way for the user through a mechanism of propagation of topological events from the topological components toward other components. Finally, the previous concepts have been extended to provide multiple topologies for the same degrees of freedom. Examples of cataract surgery simulation based on this versatile design are shown.

Keywords: Mesh Topology, Topological Change, Real-Time Simulation, User-defined Data Structure.

1   Introduction

While mesh geometry describes where mesh vertices are located in space, mesh topology tells how vertices are connected to each other by edges, triangles or any type of mesh element. Both information are required on a computational mesh to perform mesh visualization, mechanical modeling, collision detection, haptic rendering, scalar or vectorial field description.

Since topological changes are essential for surgery simulators, a common difficulty when designing those simulators is to ensure that the visual, mechanical, haptic and collision behavior of all meshes stay valid and consistent upon any topological change.

This paper proposes a framework to handle topological changes in a modular and versatile way. Our approach is modular since each software component (collision detection, mechanical solver...) may be written with little knowledge about the nature of other components. It is versatile because any type of topological changes can be handled with the proposed design.

F. Bello, E. Edwards (Eds.): ISBMS 2008, LNCS 5104, pp. 147–156, 2008.

The proposed framework has been implemented in SOFA [4], which is an Open source C++ platform for the real-time modeling of deformable structures, particularly for medical simulation and planning. While there exists several other simulation platforms, to the best of our knowledge most of them do not explicitly handle topological changes [3,2] or do so for only a limited type of meshes or mechanical behaviors [5].

Our objective to keep a modular design implies that mesh related information (such as mechanical or visual properties) is not centralized in the mesh data structure (as done using the CGAL [1] library) but is stored in the software components that are using this information. Furthermore, we manage an efficient and direct storage of information into arrays despite the renumbering of elements that occur during topological changes.

The SOFA framework is based on the integration of multiple simulation components that are associated to one or several objects. These components include *Solvers, Degrees of Freedom* (containing vertex positions), *Mechanical Force Fields, Mappings, Mass* or *Constraints*. They are considered as the leaves of a simulation tree, which describes the scene and whose internal nodes separate hierarchical levels of modeling the scene objects (*e.g. Behavior Model, Collision Model, Visual Model*). Fig. 1 shows one example of a simulation tree associated to a SOFA scene with a cylinder mesh. Note that some components may interact together and need to be synchronized. Information flow is carried on by visitors that traverse the simulation tree to propagate spatial positions top down and forces bottom up.

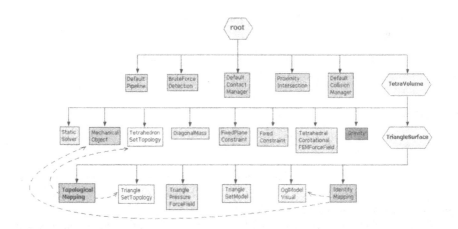

Fig. 1. Simulation Tree associated to a SOFA scene with a cylinder mesh: the cylinder is both modeled as a tetrahedral volume (2nd level) and as a triangular surface (3rd level). The lower model corresponds to the visible part of the mesh. The *Mechanical Object* component contains the mesh *Degrees of Freedom*. The dotted arrows indicates dependencies for two types of *Mapping* components, the *Topological Mapping* will be presented in Section 5.

2   Family of Topologies

We focus the topology description on meshes that are cellular complexes made of
k-simplices (triangulations, tetrahedralisation) or k-cubes (quad or hexahedron
meshes). These meshes are the most commonly used in real-time surgery simulation and can be hierarchically decomposed into k-cells, edges being 1-cells,
triangles and quads being 2-cells, tetrahedron and hexahedron being 3-cells. To
take advantage of this feature, the different mesh topologies are structured as
a family tree (see Fig. 2) where children topologies are made of their parent
topology. This hierarchy makes the design of simulation components very versatile since a component working on a given mesh topology type will also work on
derived types. For instance a spring-mass mechanical component only requires
the knowledge of a list of edges (an *Edge Set Topology* as described in Fig. 2)
to be effective. With the proposed design, this component can be used with no
changes on triangulation or hexahedral meshes.

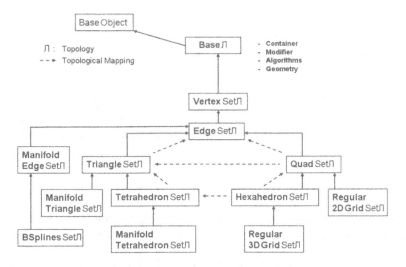

Fig. 2. Family tree of topology objects. Dashed arrows indicate possible *Topological
Mappings* from a topology object to another (see Section 5)

The proposed hierarchy makes also a distinction between conformal and manifold meshes. While most common FEM components require a mesh to be conformal (but not necessarily manifold), many high-level software components (such
as cutting, contact, haptic feedback algorithms) require the mesh to be a manifold where a surface normal is well-defined at each vertex.

Topology objects are composed of four functional members: *Container, Modifier, Algorithms* and *Geometry*. The *Container* member creates and updates
when needed two complementary arrays (see Fig. 3). The former describes the
l-cells included in a single k-cell, $l < k$, while the latter gives the k-cells adjacent to a single l-cell. The *Modifier* member provides low-level methods that

implement elementary topological changes such as the removal or addition of an element. The *Algorithms* member provides high-level topological modification methods (cutting, refinement) which decompose complex tasks into low-level ones (see Section 4). The *Geometry* member provides geometrical information about the mesh (*e.g.* length, normal, curvature, ...) and requires the knowledge of the vertex positions stored in the *Degrees of Freedom* component.

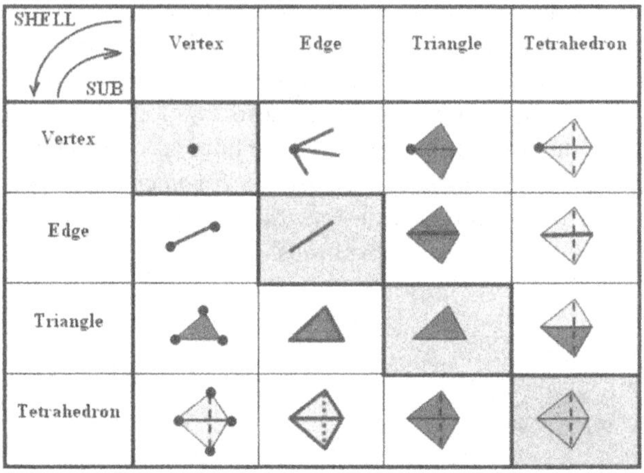

Fig. 3. The two topological arrays stored in a *Container* correspond to the upper and lower triangular entries of this table. The upper entries provide the k-cells adjacent to a l-cell, $l < k$. The lower entries describe the l-cells included in a k-cell. Similar table exists for quad and hexahedron elements.

3   Component-Related Data Structure

A key feature of our design is that containers storing mesh information (material stiffness, list of fixed vertices, nodal masses, ...) are stored in *components* and spread out in the simulation tree. This modular approach is in sharp contrast with a centralized storage of information in the mesh data structure through the use of generic pointers or template classes.

Another choice is that most containers are simple arrays with contiguous memory storage and a short direct access time. This is important for real-time simulation, but bears some drawbacks when elements of these arrays are being removed since it entails the renumbering of elements. For instance, when a single element is removed, the last array element is renumbered such that the array stays contiguous. Fortunately, all renumbering tasks that maintain consistent arrays can be automated and hidden to the user when topological changes in the mesh arise (see Section 4). Besides, time to update data structures does not depends on the total number of mesh elements but only on the number of modified elements. Therefore, in our framework, mesh data structures are stored in simple and efficient containers, the complexity of keeping the container consistent with topological changes being automated.

There are as many containers as topological elements: vertices, edges, triangles, ... These containers are similar to the STL *std::vector* classes and allow one to store any component-related data structure. A typical implementation of spring-mass models would use an edge container that stores for each edge, the spring stiffness and damping value, the ith element of that container being implicitly associated with the ith edge of the topology. Finally, two other types of containers may be used when needed. The former stores a data structure for a subset of topological elements (for instance pressure on surface triangles in a tetrahedralisation) while the latter stores only a subset of element indices.

4   Handling Topological Changes

Surgery simulation involves complex topological changes on meshes, for example when cutting a surface along a line segment, or when locally refining a volume before removing some tissue. However, one can always decompose these complex changes into a sequence of elementary operations, such as adding an element, removing an element, renumbering a list of elements or modifying a vertex position.

Our approach to handle topological changes makes the update of data structures transparent to the user, through a mechanism of propagation of topological events. A topological event corresponds to the intent to add or to remove a list of topological elements. But the removal of elements cannot be tackled in the same way as the addition of elements. Indeed, the element removal event must be first notified to the other components before the element is actually removed by the *Modifier*. Conversely, element addition is first processed by the *Modifier* and then element addition event is notified to other components. Besides, the events notifying the creation of elements also include a list of ancestor elements. Therefore, when splitting one triangle into two sub-triangles (see Fig. 5), each component-related information (*e.g.* its Young modulus or its mass density) associated with a sub-triangle will be created knowing that the sub-triangle originates from a specific triangle. Such mechanism is important to deal with meshes with non-homogeneous characteristics related to the presence of pathologies.

The mechanism to handle topological changes is illustrated by Fig. 4. The notification step consists in accumulating the sequence of topological events involved in a high-level topological change into a buffer stored in the *Container*. Then the event list is propagated to all its neighbors and leaves beneath by using a visitor mechanism, called a *Topology Visitor*. Once a given component is visited, the topological events are actually processed one by one and the data structure used to store mesh related information (see Section 3) are automatically updated.

In practice, for each specific component (*e.g.* spring-mass mechanical component), a set of callback functions are provided describing how to update the data structure (*e.g.* spring stiffness and damping values) when adding or removing an element (*e.g.* the edges defined by the two extremities of the springs). We applied the observer design pattern so that component-related data structures update themselves automatically.

Triangle Set Topology
Container
buffer
Modifier
Algorithms
Geometry
Propagate
down the simulation tree
Buffer
topological
events
Process
element
adding or
removing
Decompose
complex changes
into simple ones
Request
geometrical information
(position, normal, ...)
Component
Container
Information
Update
Information

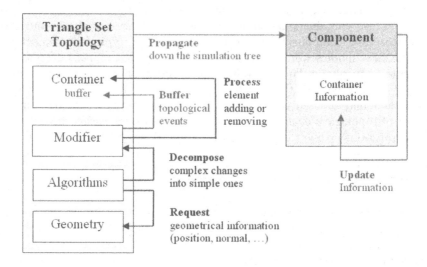

Fig. 4. Handling topological changes, with the example of a *Triangle Set Topology*. *Component* corresponds to any component of the simulation which may need topological information to perform a specific task. Black features indicate the effective change process. Red features show the steps of event notification.

Cutting algorithm in 7 steps
1. Add 4 Vertices : (e, f, g, h), defining
(e, f) defined by (b, c, 0.6) and (g, h) by (d, c, 0.4)
2. Buffer 4 Vertices Adding Event : (e, f, g, h)
3. Add 5 Triangles :
((b, e, a), (f, c, a), (b, d, g), (b, g, e), (h, c, f))
4. Buffer 5 Triangles Adding Event :
((b, e, a), (f, c, a), (b, d, g), (b, g, e), (h, c, f))
5. Buffer 2 Triangles Removing Event : ((a, b, c), (b, d, c))
6. Propagate and Handle buffered Events
7. Remove 2 Triangles : ((a, b, c), (b, d, c))
a
b
c
d
e
f
g
h

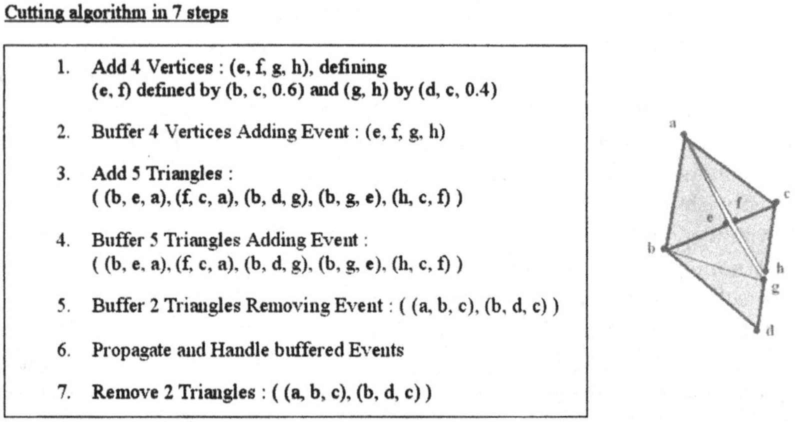

Fig. 5. Seven steps to perform cutting along two triangles (these generic steps would be the same to cut an arbitrarily number of triangles). Black steps indicate the effective change process. Red steps show the steps of event notification.

5 Combining Topologies

Handling a single mesh topology in a surgery simulation scene is often too restrictive. There are at least three common situations where it is necessary to have, for the same mesh, several topological descriptions sharing the same degrees of freedom: the *boundary, composite* and *surrogate* cases. In the *boundary* scenario, specific algorithms may be applied to the boundary of a mesh, the boundary of tetrahedral mesh being a triangulated mesh and that of a triangular mesh being a polygonal line. For instance, those algorithms may consist of applying additional membrane forces (*e.g.* to simulate the effect of the Glisson capsule in the liver) or visualizing a textured surface. Rather than designing specific simulation components to handle triangulations as the border of tetrahedrisations, our framework allows us to create a triangulation topology object from a tetrahedrisation mesh and to use regular components associated with triangulations.

The *composite* scenario consists in having a mesh that includes several types of elements: triangles with quads, or hexahedra with tetrahedra. Instead of designing specific components for those composite meshes, it is simpler and more versatile to reuse components that are dedicated to each mesh type. Finally the *surrogate* scenario corresponds to cases where one topological element may be replaced by a set of elements of a different type with the same degrees of freedom. For instance a quad may be split into two triangles while an hexahedron may be split into several tetrahedra. Thus a quad mesh may also be viewed as a triangular mesh whose topology is constrained by the quad mesh topology.

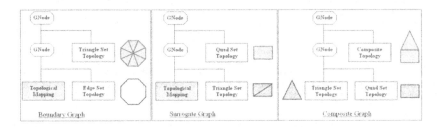

Fig. 6. Three scenari examples to combine topologies. (*From left to right*) A *Boundary Graph* from a triangle set to an edge set, a *Surrogate Graph* from a quad set to an triangle set and a *Composite Graph* superseding a quad set and a triangle set.

These three cases can be handled seamlessly by using a graph of multiple topologies, the topology object being at the top node having the specific role of controlling the topologies below. Although we chose to duplicate topological information in memory, it has no effect on the time required to compute the forces. Fig. 1 provides an example of a border scenario (triangulation as the border of a tetrahedralisation) while Fig. 6 shows general layout of topology graphs in the three cases described previously. In any cases, those graphs include a dedicated component called a *Topological Mapping* whose objectives are twofold. First, they translate topological events originating from the master topology (*e.g.* remove this quad) into topological actions suitable for the slave topology (*e.g* remove those two triangles for a surrogate scenario). Second, they

Fig. 7. SOFA scene to simulate the cataract surgery: the patient's eye is held open by a lid speculum to be easily accessed

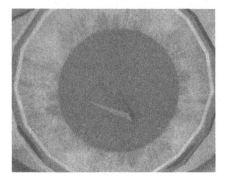 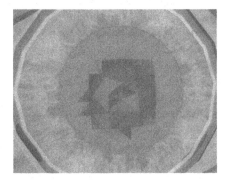

Fig. 8. (*Left*) The upper triangular surface modeling the capsule is cut and deforms as a membrane. (*Right*) Once the capsule is cut and teared off, some tetrahedral volume of the lens is removed.

provide index equivalence between global numbering of elements in the master topology (*e.g* a triangle index in a tetrahedralisation topology) and local numbering in the slave topology (*e.g* the index of the same triangle in the border triangulation). Possible *Topological Mappings* from a topology object to another have been represented by blue arrows in Fig. 2.

Note that those topology graphs can be combined and cascaded, for instance by constructing the triangulation border of a tetrahedrisation created from an hexahedral mesh. But only topology algorithms of the master topology may be called to simulate cutting or to locally refine a volume. By combining topology graphs with generic components one can simulate fairly complex simulation scenes where topological changes can be seamlessly applied.

6   Results

The contribution of our design as the topology module in the SOFA project lead to the development of a clinical application: the simulation of cataract surgery.

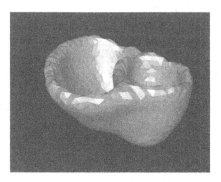
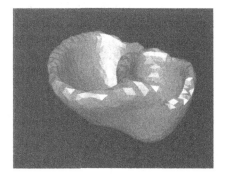

Fig. 9. (*Left*) The heart is composed of 16553 tetrahedral elements. (*Right*) 557 tetrahedra included in a given sphere were removed within a time of 196 milliseconds.

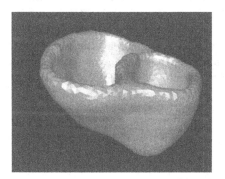
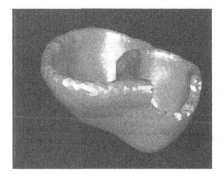

Fig. 10. (*Left*) The heart is composed of 97087 tetrahedral elements. (*Right*) 2455 tetrahedra included in a given sphere were removed within a time of 883 milliseconds.

Cataract is an opacity that develops in the natural lens of the eye, causing a progressive vision loss. Its surgery consists in first opening the capsule that contains the lens by making an incision (i.e. Capsulorrhexis step), then in extracting the lens (i.e. Phacoemulsification step). After lens removal, the surgeon inserts an artificial intraocular lens through the incision of the membrane. All numerical results were obtained on a Pentium-M at 2.13 Ghz with 1.00 Gb RAM.

Fig. 7 and Fig. 8 shows the simulated steps in a SOFA scene, where we chose to apply a Membrane and Bending FEM Model on a triangular mesh for the capsule, and a Corotational FEM Model [10] on a tetrahedral mesh for the lens. The eye capsule is composed of 760 triangular elements; to perform its first incision, we applied the seven steps algorithm (see Fig. 5) which deleted 22 ancestor triangles and created 64 new triangles, with a cutting time of 78 milliseconds. The eye lens is composed of 5484 tetrahedral elements; 166 tetrahedra included in a given sphere were removed within a time of 62 milliseconds.

Other results are presented on Fig. 9 and Fig. 10 for the simulation of heart surgery in SOFA. We applied the Corotational FEM Model on two meshes of the same heart with different resolution, from which an infarct region of the left ventricule is removed. Note that the ratio between the removal time and the number of removed tetrahedra corresponds to the average time to remove one tetrahedron, which is around 0.35 milliseconds. Thus, the required time to remove a set of tetrahedra while updating topological information is proportional to the number of tetrahedra to remove.

7   Conclusion and Perspectives

To conclude, our approach handles topological changes in an innovative way, by storing information into arrays to ensure a short memory access time. Besides, mesh related data is not centralized in a common data structure but may be defined by the user in several dedicated components acting on the mesh. The cost to update the designed data structures is proportional to the number of modified elements, independently of the total number of mesh elements, hence the ability to perform local changes for simulation under real-time constraints.

Future work will include the design and implementation of more complex cutting and refinement tasks on surface and volumetric meshes and will also extend the available topologies to include finite elements of high orders.

References

1. Cgal : Computational geometry algorithms library, http://www.cgal.org
2. Gipsi : General physical simulation interface, http://gipsi.case.edu
3. Opentissue, http://www.opentissue.org
4. Sofa : Simulation open framework architecture, http://www.sofa-framework.org
5. Springs, http://spring.stanford.edu
6. Agus, M., Gobbetti, E., Pintore, G., Zanetti, G., Zorcolo, A.: Real-time cataract surgery simulation for training. In: Eurographics Italian Chapter Conference, Catania, Italy, Eurographics Association (2006)
7. Allard, J., Cotin, S., Faure, F., Bensoussan, P.-J., Poyer, F., Duriez, C., Delingette, H., Grisoni, L.: Sofa – an open source framework for medical simulation. In: Medicine Meets Virtual Reality (MMVR 1915), Long Beach, USA (February 2007)
8. Forest, C., Delingette, H., Ayache, N.: Removing tetrahedra from manifold tetrahedralisation: application to real-time surgical simulation. Medical Image Analysis 9(2), 113–122 (2005)
9. Molino, N., Bao, Z., Fedkiw, R.: A virtual node algorithm for changing mesh topology during simulation. ACM Transactions on Graphics 23(3), 385–392 (2004)
10. Nesme, M., Payan, Y., Faure, F.: Efficient, physically plausible finite elements. In: Dingliana, J., Ganovelli, F. (eds.) Eurographics (short papers) (August 2005)

Contact Model for Haptic Medical Simulations

Guillaume Saupin[1], Christian Duriez[2], and Stephane Cotin[2]

[1] CEA, Fontenay aux Roses, France
[2] INRIA, University of Lille, France

Abstract. In surgery simulation, precise contact modeling is essential to obtain both realistic behavior and convincing haptic feedback. When instruments create deformations on soft tissues, they modify the boundary conditions of the models and will mainly modify their behavior. Yet, most recent work has focused on the more precise modeling of soft tissues while improving efficiency; but this effort is ruined if boundary conditions are ill-defined. In this paper, we propose a novel and very efficient approach for precise computation of the interaction between organs and instruments. The method includes an estimation of the contact compliance of the concerned zones of the organ and of the instrument. This compliance is put in a buffer and is the heart of the multithreaded local model used for haptics. Contact computation is then performed in both simulation and haptic loops. It follows unilateral formulation and allows realistic interactions on non-linear models simulated with stable implicit scheme of time integration. An iterative solver, initialized with the solution found in the simulation, allows for fast computation in the haptic loop. We obtain realistic and physical results for the simulation and stable haptic rendering.

1 Introduction

Medical training methods based on synthetic or animal tissues do not provide a satisfactory solution since they cannot represent the complexities of real-life anatomy and pathological situations. Computer-based medical training systems offer a very elegant solution to this problem since it can theoretically provide a realistic and configurable training environment, assuming that the core simulation of the system is sufficiently precise and fast.

Indeed, this simulation must be able to reproduce the physical behavior of the anatomy during the intervention and must adapt the computation, in real-time to the interactions with the virtual surgical instrument. In addition, the equipment includes a user interface for acquiring the gesture. The immersion during the training session is improved by adding the sense of touch: the user interface can provide an haptic feedback on the simulation.

However, if current available solutions propose realistic physical models of the organs, they often oversimplify the models of contact. Yet, most medical

---

[1] guillaume.saupin@cea.fr

F. Bello, E. Edwards (Eds.): ISBMS 2008, LNCS 5104, pp. 157–165, 2008.
© Springer-Verlag Berlin Heidelberg 2008

procedures rely on interactions between a device and the anatomy, therefore involving many instances of complex contacts. Current approaches often rely on a simple contact point (see for instance [3]), rather than an actual area of contact, and rarely account for friction. While this simplification can produce plausible results in the case of an interaction between the end effector of a surgical instrument and the surface of an organ, it is generally an incorrect approximation.

Contributions: In this paper we propose a novel and very efficient approach for precise computation of contact response in both simulation and haptic control loops. We propose a formalism where a contact model can be derived from the behavior model of the organs while verifying physically based contact and friction laws on multiple coupled contacts. Starting from a visco-elastic deformation law of the tissues, we introduce a pre-computional step for the on-line approximation of a compliance matrix that is used during contact process. This matrix, computed by the simulation, is the core model of the haptic control loop which is processed in a separate thread and updated at 1kHz. The method guarantees that all interpenetrations are solved at the end of each time step in the simulation while providing stable and efficient haptic feedback.

2   Previous Work

Contact modeling in medical simulation is a challenging problem. The way contacts are handled plays a very important role in the overall behavior of the interacting objects. The choice of the contact model and the inclusion or not of friction, highly influence the post-impact motion of the interacting objects. Additionally, when a contact between objects occurs, it induces quick changes in the haptic force which can cause artificial growth of energy and can lead to instabilities.

In most simulators with haptic feedback, the collision response of soft tissues with a virtual surgical instrument is assumed to be very local: the interaction only consider a single point [11] [10]. The most popular approach is the penalty method which consists in defining a contact force $f = k\delta$ at each contact point where δ is a measure of the interpenetration between a pair of colliding objects, and k is a stiffness parameter. This stiffness parameter must be large enough to avoid any visible interpenetration, however, its value cannot be determined exactly. In addition, if an explicit time integration scheme is used, and k is large, very small time steps are required to guarantee the stability. The quick growth of energy in the haptic control loop induced by the method often leads to excessive damping in the provided solutions. A possible improvement over the penalty method can be achieved through the use of an implicit integration scheme [12]. Yet, solving the resulting stiff and non-smooth system can be computationally prohibitive when the objective is to reduce as much as possible the interpenetration distance.

Some methods, developed for force feedback applications are based on the avoidance of visible interpenetration through constraint-based techniques. The collision can be prevented by geometrical deformation constraints like in [16] or [6], or by god-object [17] and proxy [4] methods. Another way is the use of Lagrange multipliers, which are appropriate for handling bilateral constraints [7]. However, contacts between objects intrinsically define unilateral constraints, which means that physics is not always verified when using techniques based on Lagrange multipliers. As a consequence, colliding objects could stay artificially stuck at the end of the time step.

Improvements over constraint-based techniques are possible by using a Linear Complementary Problem (LCP) formulation. The solution of the LCP gives an accurate description of the contact forces needed to zero out the interpenetration, and prevents objects to stick together [15]. By expanding the LCP, or by using a non-linear solver, the formulation can be extended to model both static and dynamic friction for rigid [2] and deformable [5] objects. Computationally efficient methods for solving linear complementary problems are proposed [14], thus making such approaches appealing even for interactive simulations. Yet, when dealing with haptic feedback, real-time computation of the solution is almost impossible since the LCP algorithm relies on the computation of the inverse of the stiffness matrix for each object in contact. This inverse can be exactly pre-computed in the case of linear elastic models [5]. In this paper, we present a method to estimate this inverse, through a compliance, for non-linear deformable models of the soft tissues.

3   Simulation

In this section we present the main steps involved in interactive medical simulation. Soft tissues are often modeled using visco-elastic deformation laws, and usually includes dynamics. Mathematical equations that are obtained have led to a synthetic formulation:

$$\mathbf{M}\dot{\mathbf{v}} = \mathbf{p} - \mathbb{F}\left(\mathbf{q}, \mathbf{v}\right) + \mathbf{r} \qquad (1)$$

where $\mathbf{q} \in \mathbb{R}^n$ is the vector of generalized degrees of freedom, $\mathbf{M}$ is the mass matrix, $\mathbf{v} \in \mathbb{R}^n$ is the vector of velocity. $\mathbb{F}$ represents internal visco-elastic forces, and $\mathbf{p}$ gathers external forces. $\mathbf{r} \in \mathbb{R}^n$ is the vector of contact forces contribution.

In the following of the section, the processing of this equation through the simulation is described. Then we focus on the computation of the contact forces $\mathbf{r}$ and its impact on the computed motions.

3.1   Simulation Overview

Our simulation entails the following steps:

1. The integration of soft tissues models without considering contact forces and the update of the instrument position. This step is called the *free motion*.
2. The *collision detection* algorithm determines which points of the tissue surface are in contact with the surgical instrument.

3. The computation of a *collision response* that follows the contact and friction laws at each contacting point.
4. Finally, the displacement field induced by contact forces is applied during the *constrained motion*

The *free motion* computation is based on the implicit Euler integration scheme of all forces except the contact forces:

$$\mathbf{M}\dot{\mathbf{v}}_f = \mathbf{p} - \mathbb{F}(\mathbf{q}_f, \mathbf{v}_f) \tag{2}$$

where $\mathbf{q}_f$ and $\mathbf{v}_f$ are positions and velocities at the free motion, that are unknown. We face a non-linear problem.

To deal with the problem, we use the position and velocities values at the beginning of the time step: $\mathbf{q}_i$ and $\mathbf{v}_i$ which are known from the previous time step. We can introduce a unique unknown, $d\mathbf{v}$ using the following relations:

$$\dot{\mathbf{v}}_f = \frac{d\mathbf{v}}{h} \tag{3}$$

$$\mathbf{v}_f = \mathbf{v}_i + d\mathbf{v} \tag{4}$$

$$\mathbf{q}_f = \mathbf{q}_i + h\mathbf{v}_f = \mathbf{q}_i + h\mathbf{v}_i + hd\mathbf{v} \tag{5}$$

where h is the time step. Then, we introduce a linearization of the internal forces:

$$\mathbb{F}(\mathbf{q} + hd\mathbf{v}, \mathbf{v} + d\mathbf{v}) \approx \mathbb{F}(\mathbf{q}, \mathbf{v}) + \frac{d\mathbb{F}}{d\mathbf{q}}hd\mathbf{v} + \frac{d\mathbb{F}}{d\mathbf{v}}d\mathbf{v} \tag{6}$$

Finally, at each time step, we compute the motion by solving the following system:

$$\left(\frac{1}{h}\mathbf{M} + \mathbf{B} + h\mathbf{K}\right)d\mathbf{v} = -\mathbb{F}(\mathbf{q}_i + h\mathbf{v}_i, \mathbf{v}_i) + \mathbf{p} \tag{7}$$

where $\mathbf{B} = \frac{d\mathbb{F}}{d\mathbf{v}}$ and $\mathbf{K} = \frac{d\mathbb{F}}{d\mathbf{q}}$ are the damping and stiffness (tangent) matrices.

The *collision detection* detection must provide a set of colliding points between the surfaces of the virtual instrument and the soft tissue model. In our case, we have extend the method presented in [9] to deformable objects. At the end of steps 1 and 2, the objects has moved freely and may interpenetrate. During steps 3 and 4 we evaluate the contact forces $\mathbf{r}$ and apply theses forces to ensure non-interpenetration. This steps are based on the displacement correction observed during contact processing:

$$\left(\frac{\mathbf{M}}{h^2} + \frac{\mathbf{B}}{h} + \mathbf{K}\right)d\mathbf{q} = \mathbf{r} \tag{8}$$

3.2   Contact Force Computation

Once the free motion and the collision detection has been performed, the simulation is provided with a vector of interpenetration distances, called δ_{free}. In this section, we will show how to compute the contact forces avoiding interpenetration.

LCP Formulation. The points where two objects are colliding usually does not correspond directly to one of the degrees of freedom. Therefore, we need to define the operator $\mathbf{H}$ which maps the displacements of the model $d\mathbf{q}$ in the global frame to the relative displacement used in the contact laws:

$$\mathbf{H}d\mathbf{q} = \boldsymbol{\delta} - \boldsymbol{\delta}^{free} \tag{9}$$

Virtual work principle provides the transposed relation for the contact forces $\mathbf{f}$:

$$\mathbf{r} = \mathbf{H}^T\mathbf{f} \tag{10}$$

Then, to respect the frictionless contact case, we build and solve the following Linear Complementarity Problem (LCP):

$$\begin{cases} \boldsymbol{\delta} = \underbrace{\mathbf{HCH}^T}_{\mathbf{W}}\mathbf{f} + \boldsymbol{\delta}^{free} \\ 0 < \boldsymbol{\delta} \perp \mathbf{f} > 0 \end{cases} \tag{11}$$

Where $\mathbf{C} = \left(\frac{1}{h^2}\mathbf{M} + \frac{1}{h}\mathbf{B} + \mathbf{K}\right)^{-1}$ is called the compliance matrix. We use an extended formulation of this LCP in order to include friction computation and solve the LCP using an iterative, gauss-seidel like solution, like in [5].

Contact Compliance Warping. The computation of operator $\mathbf{W}$ necessitates the compliance matrix of the colliding objects. With deformable objects undergoing large displacement, $\mathbf{C}$ must be updated when $\mathbf{M}$, $\mathbf{K}$ and $\mathbf{B}$ change according to the non-linear deformations. This computation is highly time-consuming. That is why we propose to evaluate the compliance using its value in the rest shape configuration $\mathbf{C}^0$. To account for the non-linear rotation induced by the deformation, we propose to evaluate a rotation at each node, from its rest shape to its current shape. This evaluation could be done geometrically or based on gradient measure [8]. We obtain an approximate value of the compliance $\tilde{\mathbf{C}}$:

$$\tilde{\mathbf{C}} = \mathbf{RC}^0\mathbf{R}^T \tag{12}$$

Where $\mathbf{R}$ is a 3×3 block diagonal matrix that gathers the rotations associated with object nodes. This simplification significantly speeds up the computation because $\mathbf{C}^0$ can be precomputed.

3.3   Constrained Motion Integration

When building $\mathbf{W}$, we use an approximate compliance. Therefore, to get a motion that is coherent with the LCP, we compute the constrained motion using the same *compliance*. The exact steps involved in the constrained motion follows:

1. We map the contact forces in the rest shape coordinate frame: $\mathbf{r}^0 = (\mathbf{HR})^T\mathbf{f}$
2. We compute the displacement in the original coordinate: $d\mathbf{q}^0 = \mathbf{C}^0\mathbf{r}^0$
3. We rotate back the displacement to the current coordinate frame: $d\mathbf{q} = \mathbf{R}d\mathbf{q}^0$

Nevertheless, as a consequence, the corrective motion is not based on the exact compliance of the tissue model. However, contact and friction laws are followed and the obtained motions avoid disturbing interpenetration. Moreover, the approximation is partly corrected by the free motion, based on exact constitutive law, of the next time-step.

4   Haptic Coupling

In order to enhance haptic rendering stability and to compute contact forces, we extend the god-object principle introduced in [17] to deformable models and friction. We separate the haptic loop from the simulation loop so that satisfying force feedback is achieved.

4.1   Haptic Device and God Object Coupling

God-object has been defined as the answer of the following observation: we can not prevent the points associated with the haptic device from penetrating simulation's objects. However, we can attached a virtual object, subject to the physical laws of the simulation, to the haptic device. Thus, this virtual object, the god-object, can not penetrate other objects, and occupies the position that the haptic object would have had. Moreover, force feedback can be easily computed knowing haptic device and god-object positions.

The haptic device and the god-object are coupled using simple impedance control techniques. Their coupling is governed by the equation below:

$$\mathbf{M}\dot{\mathbf{v}} = \mathbb{F}_c\left(\mathbf{q}, \mathbf{q}_h, \mathbf{v}, \mathbf{v}_h\right) + \mathbf{r} \tag{13}$$

Where $\mathbf{M}$ is the mass of the surgical instrument driven by the haptic device ; $\mathbb{F}_c$ represents the forces computed in the control loop of the haptic device ; $\mathbf{q}$ and $\mathbf{v}$ are the position and velocity of the god-object ; $\mathbf{q}_h$ and $\mathbf{v}_h$ are the position and velocity measured on the device; $\mathbf{r}$ is the opposite of the contact force presented in equation (1).

In our simulation we simply use a proportional derivative control which can be rewriten:

$$\mathbb{F}_c\left(\mathbf{q}, \mathbf{q}_h, \mathbf{v}, \mathbf{v}_h\right) = \mathbf{K}_c(\mathbf{q} - \mathbf{q}_h) + \mathbf{B}_c(\mathbf{v} - \mathbf{v}_h)$$

The simulation of the god object will follow the same scheme than the other simulated object. A *free motion* will integrate the motion based on the provided values of $\mathbf{q}_h$ and $\mathbf{v}_h$, and without the contact forces. Then, we will compute and inverse the matrix $\left(\frac{\mathbf{M}}{h^2} + \frac{\mathbf{B}_c}{h} + \mathbf{K}_c\right)$ to add the contribution of the god-object in the LCP compliance matrix. With the value of the contact forces $\mathbf{r}$, we compute the constrained motion of the god-object.

4.2   Multifrequence Simulation

In this work, we propose to separate the simulation from the haptic rendering, as it has been already proposed many times (for instance, see [13] [16] [6] [17]). The

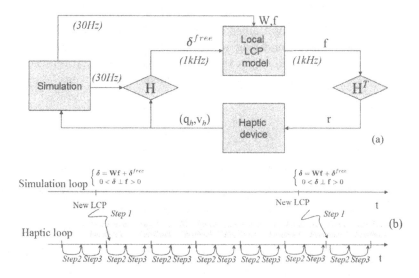

Fig. 1. (a) Coupling Scheme (b) Time line of the two loops. *Step 1*: **W**, **f** are shared, *Step 2*: δ^{free} is updated, *Step 3*: Re-solve the LCP (for a new **f**) using previous value as an initial guess.

originality of our approach lies in the the use of the compliance value of the LCP matrix as a local model. Two LCPs will be simultaneously solved in the simulation and in the haptic loop. The one in the simulation has already been presented. When it is solved, it is shared with the haptic loop (*Step 1*). In the LCP of the haptic loop, the value of δ^{free} is updated using the new position from the device (*Step 2*). We use the solution from the simulation as an initial guess for the iterative solver of the haptic loop LCP (*Step 3*). With the time coherency, the initial guess is close to the good solution, so it converges quickly. *Step 2* and *Step 3* are repeated continuously in the haptic loop and the contact force computed in the LCP is applied as a new force feedback at 1kHz. Meanwhile, the simulation loop performs its calculation, but a at a lower frequency.

5   Examples

In the examples below, we use a corotational model to simulate the non-linear deformation of a liver. The simulation is run on a Intel bi-dual core 5130 @2Ghz with 2 Giga RAM. A visco-elastic behavior is obtained thanks to the use of Rayleigh coefficients for damping model.

   In this first benchmark, the liver model contains about 1300 tetrahedra. We use Dirichlet conditions on some nodes to fix the model in the body. A better model of the boundary conditions would require to account for the anatomical data of the ligaments and the blood vessel network. In the presented scenario, the liver model is manipulated with a grasper. Since our model handles friction, we can catch a part of the virtual liver using this grasper.

We handle up to 150 contact points, i.e. 450 constraints while keeping a simulation framerate of about 50 frames/sec. With a computation time of 20ms for one step, a direct haptic rendering would not be possible. However, since we have separated the simulation from the force feedback computation, the haptic rendering is very satisfying as it works at a frequency near 1 Khz.

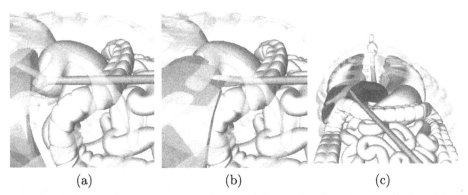

(a)   (b)   (c)

Fig. 2. The liver and instrument interactions. (a)the instrument just touch the surface (b) the instrument grasp the liver (c) Global view of the simulation.

6   Conclusion and Future Work

This paper shows how satisfying haptic rendering can be achieved in physically based simulations of highly deformable objects involving friction with surgery instruments. Especially we explain how this can be done using contact correction approximation which nonetheless provide physically plausible contact forces. We also show how to compute force feedback in a high frequency separate thread using the LCP of the previous simulation step.

Future work will concentrate on improving the method on a more realistic scenario, in particular with the modeling of contact between soft tissues, and with more realistic deformable models.

Acknowledgment

This work was made possible by the SOFA framework [1] (www.sofa-framework. org).

References

1. Allard, J., Cotin, S., Faure, F., Bensoussan, P.-J., Poyer, F., Duriez, C., Delingette, H., Grisoni, L.: Sofa an open source framework for medical simulation. In: Medicine Meets Virtual Reality (MMVR 15), Long Beach, USA (February 2007)
2. Anitescu, M., Potra, F., Stewart, D.: Time-stepping for three-dimentional rigid body dynamics. Computer Methods in Applied Mechanics and Engineering (177), 183–197 (1999)

3. Balaniuk, R.: A differential method for the haptic rendering of deformable objects. In: VRST 2006: Proceedings of the ACM symposium on Virtual reality software and technology, pp. 297–304. ACM, New York (2006)
4. Barbagli, F., Salisbury, K., Prattichizzo, D.: Dynamic local models for stable multi-contact haptic interaction with deformable objects. Haptic Interfaces for Virtual Environment and Teleoperator Systems 2003, 109–116 (2003)
5. Duriez, C., Dubois, F., Kheddar, A., Andriot, C.: Realistic haptic rendering of interacting deformable objects in virtual environments. IEEE Transactions on Visualization and Computer Graphics 12(1), 36–47 (2006)
6. Forest, C., Delingette, H., Ayache, N.: Surface contact and reaction force models for laparoscopic simulation. In: International Symposium on Medical Simulation (June 2004)
7. Galoppo, N., Otaduy, M.A., Mecklenburg, P., Gross, M., Lin, M.C.: Fast simulation of deformable models in contact using dynamic deformation textures. In: SCA 2006, Switzerland, pp. 73–82. Eurographics Association (2006)
8. Hauth, M., Straßer, W.: Corotational simulation of deformable solids. In: WSCG 2004, pp. 137–145 (2004)
9. Johnson, D., Willemsen, P., Cohen, E.: Six degree-of-freedom haptic rendering using spatialized normal cone search. IEEE Transactions on Visualization and Computer Graphics 11(6), 661–670 (2005)
10. Mahvash, M., Hayward, V.: High-fidelity haptic synthesis of contact with deformable bodies. IEEE Computer Graphics and Applications 24(2), 48–55 (2004)
11. Mendoza, C., Sundaraj, K., Laugier, C.: Faithfull force feedback in medical simulators. In: International Symposium in Experimental Robotics, vol. 8. Springer, Heidelberg (2002)
12. Meseure, P.: A physically based virtual environment dedicated to surgical simulation. In: Surgery Simulation and Soft Tissue Modeling (IS4TM), pp. 38–47 (June 2003)
13. Morin, S., Redon, S.: A force-feedback algorithm for adaptive articulated-body dynamics simulation. In: 2007 IEEE International Conference on Robotics and Automation, April 10-14, 2007, pp. 3245–3250 (2007)
14. Murty, K.: Linear Complementarity, Linear and Nonlinear Programming. Internet Edition (1997)
15. Pauly, M., Pai, D.K., Guibas, L.J.: Quasi-rigid objects in contact. In: SCA 2004, Switzerland, pp. 109–119. Eurographics Association (2004)
16. Picinbono, G., Lombardo, J.-C., Delingette, H., Ayache, N.: Improving realism of a surgery simulator: linear anisotropic elasticity, complex interactions and force extrapolation. Journal of Visualisation and Computer Animation 13(3), 147–167 (2002)
17. Zilles, C.B., Salisbury, J.K.: A constraint-based god-object method for haptic display. In: IEEE IROS 1995: Proceedings of the International Conference on Intelligent Robots and Systems, pp. 31–46 (1995)

Estimation of the Fracture Toughness
of Soft Tissue from Needle Insertion

Toufic Azar and Vincent Hayward

Center for Intelligent Machines, McGill University
Montreal, Quebec, H3A 2A7, Canada
`toufic.azar@mail.mcgill.ca`

Abstract. A fracture mechanics approach was employed to develop a
model that can predict the penetration force during quasi-static needle insertion in soft tissue. The model captures a mechanical process
where the sharp needle produces a crack that is opened to accommodate the shaft of the needle. This process involves the interchange of
energy between four distinct phenomena: the work done by the needle,
the irreversible work of fracture, the work of friction, and the change
in recoverable strain energy. From measurements made in vivo, porcine
liver fracture toughness was estimated from the difference in penetration
force between two consecutive insertions at the same location. The values obtained fall within a reasonable range and confirm the relevance of
a computational model of needle insertion based on fracture mechanics.

Keywords: Needle Force, Fracture Mechanics, Fracture Toughness.

1   Introduction

Needle insertion is extensively used in medical intervention. Its use has expanded
to a variety of minimally invasive and percutaneous procedures such as biopsies,
neurosurgery and brachytherapy [1]. For pre-intervention planning, for intra-
intervention guidance, for post-intervention assessment, as well as for simulation
purposes, a model of the physics of needle interaction with tissues during insertion would be valuable.

Here, we discuss a model that accounts for both the tissue and needle characteristics. Using in vivo measurements of needle insertion in pig liver performed
by Maurin et al. [13], we could estimate the work of fracture and determine the
liver fracture toughness. These values were found by comparing the response
of the organ to a first penetration with the response of the same organ at the
pre-cracked location. We found values that are in line with others cited in the
literature and that were obtained using different methods.

The model that we developed can be used as a basis for several types of
simulation. It can be used in a finite element analysis to find deflection from
loading. Conversely, these values can be used to predict forces as a response to
displacement for planning and simulation.

F. Bello, E. Edwards (Eds.): ISBMS 2008, LNCS 5104, pp. 166–175, 2008.

2   Related Work

Roughly speaking, needle insertion models fall under two categories. In the first category, needle penetration is viewed mostly from the view point to the deformation of tissues that the interaction causes. These models represent the effect of crack formation but do not actually model it because they do not account for the underlying physics of the insertion process. Instead, stiffness-like functions are fit to experimental data. These models cannot be invariant with respect to the choice of units. In the second category, models are based on the description of the mechanics of crack formation. The effects of a needle geometry interacting with tissue properties are responsible for the permanent changes made to an organ at the scale of cells and membranes. Despite these complexities, needle piercing and cutting create new surfaces inside a body, that is, damage. In the second category, damage is modeled by averaging over surfaces and through time. They are therefore based on energy methods.

Deformation models and function fit. Maurel represent the penetration force resulting from deformation as an exponential function of the depth [12], while Brett et al. use viscoelastic functions [5]. Simone and Okamura describe the needle force as the sum of the cutting force, the stiffness force and the friction force [15]. The stiffness-like model is equated to a second-order polynomial fitted to the data. Maurin et al. and Barbé et al. obtained fits using various functions [3,4,13]. Another example of this approach is Dimaio and Salcudean's FEM models to determine the tissue deformation [7]. Recently, Hing et al. measured internal tissue deformation during needle penetration [10]. Since these techniques cannot represent combinations of tissue types and needle geometries, each case requires the determination of a new set of arbitrary parameters.

Energy models. Based on the work of Atkins [2], a multi-phase model suitable for the prediction of cutting forces was introduced by Mahvash and Hayward using energy and fracture mechanics [11]. The force of cutting is modeled as the result of an interchange of energy between the internal strain energy stored in the body due to initial deformation, the work done by the tool and the irreversible fracture work. Cutting and piercing both create new surfaces in a body. Therefore it should be possible to apply a fracture energy approach to needle insertion. In this view, the geometry that determines the cutting force is the size of the created surfaces. One may imagine that a needle causes a simple expansion to a cylindrical cavity. Shergold and Fleck portrayed the phenomenon as a mode-I crack propagation with elastic deformation to accommodate the cylindrical shaft which suggests different crack geometries [17,18]. These geometries were observed in human skin and rubber samples. Using conservation of energy and neglecting friction, they determined the penetration pressure and crack size as a function of sharp-tipped punch diameters and tissue characteristics. Besides crack geometry, the second key parameter is tissue specific resistance to fracture. Authors have devised many types of experiments to determine it.

Testing approaches (Mode-I). Atkins used a guillotine-test to estimate the fracture toughness of skin and other soft tissue [2]. Mahvash and Hayward, robotically cutting potato and liver, as well as Doran et al., cutting animal skin, assumed that during the cutting phase the work of the tool was entirely transfered to the formation of the crack [8,11]. Pereira et al. used scissor cutting to obtain fracture toughness [16]. Chanthasopeephan et al. used scalpel cutting of liver [6]; the tool work was obtained from integration of the force-displacement curve. McCarthy et al. tackled the case of scalpel cutting where the tool is in permanent contact with the tissue [14]. Because friction was not be neglected, the work of fracture was equated to the work of the difference in cutting force between two consecutive insertions at the same location.

3   Needle Insertion Phases

Needle insertion and retraction at constant velocity experimentally suggests that the process can be decomposed into phases [11,13]. With reference to Fig. 1, four phases can be identified:

Phase 1: Deformation. The needle tip comes into contact with the tissue and deforms it without penetration. It terminates with puncture.

Phase 2: Steady-state penetration. Once an energy threshold is reached, the liver capsule ruptures and a crack initiates. Then, steady-state insertion initiates and the force increases with depth. This phase terminates when the needle stops.

Phase 3: Relaxation. When the desired depth is reached the motion is stopped and the force relaxes due to the viscoelastic properties of the material. The stored strain energy combined with friction creates a suction effect.

Phase 4: Extraction phase. The needle force is due to the friction force and the release of the stored elastic strain energy.

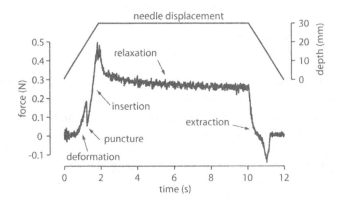

Fig. 1. Phases of needle insertion. This data, used with permission, is from [13].

3.1   Mechanism of Needle Insertion into Liver

Shergold et al. hypothethized and verified that the insertion of a sharp punch into skin and rubber causes the formation and wedging open of a mode-I crack, see Fig. 2 [18].

Fig. 2. Left: Geometry of a Franseen needle tip, forces acting on it and expected resulting 3-way branching or star-cracks. Right: standard 22° bevel tip and 45° bevel tip biopsy needles next to the expected resulting line crack.

To verify this theory, we obtained porcine liver samples directly from the slaughter house less than one day postmortem (refrigerated). The samples were manually penetrated with different needles types: (1) Franseen tip suture needle of diameter 0.84 mm; (2) sharp standard bevel tip syringe needles of diameter 0.71 mm, 1.27 mm, and 2.1 mm (22° bevel); (3) true short bevel tip liver biopsy needle (45° bevel tip) of diameter 1.47 mm. The needles were dipped in trypan blue stain before insertion in an effort to increase image contrast. The samples were prepared with thin blades and the average size of the samples was 5×5×3 mm. The cracks were examined using a binocular inverted microscope under ×10 magnification. The cracks were measured at different depths, capsule level, 3 mm depth and 6 mm depth whenever it was possible to observe them. Fig. 3 shows representative examples of the images obtained. In the figures, images were processed using Photoshop®, first discarding color information and then precisely bracketing the brightness levels to enhance contrast.

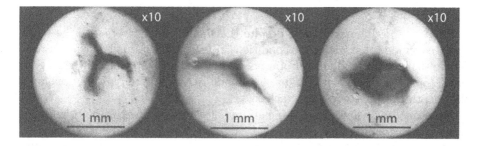

Fig. 3. Microscopic observations of fresh porcine liver penetrated with three types of needles. The left image shows the crack geometry made by a Franseen tip, the middle image show a crack resulting from a standard bevel tip, and the right image the crack left by a short bevel biopsy needle.

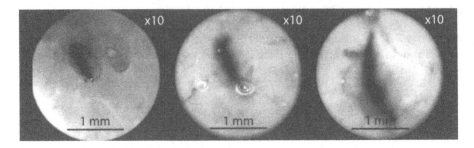

Fig. 4. Observations of fresh porcine liver penetrated with standard bevel needles of different diameters. From left to right, diameters of 0.71 mm, 1.27 mm, and 2.10 mm.

Fig. 4 shows cracks made by standard bevel needles with three different diameter. Cases where unstable crack formation was observed, that is, when the assumption of a sharp interaction breaks down were not considered. Unstable crack formation can be attributed to the non-homogeneity of the liver or unpredictable stresses induced by manual insertion.

Table 1 collects a summary of the resulting crack sizes as a function of needle geometries. Only averages as shown as the sample size is too small to allow for statistical analysis.

Table 1. Crack length as a function of needle geometry (number of observations)

needle tip	diameter (mm)	mean crack length at capsule level (mm)	standard deviation (nb. observed)	additional observations (nb. observed) at 3 mm	at 6 mm
Franseen	0.84	1.32	0.10 (9)	1.25 (6)	1.2 (3)
22° bevel	0.71	0.70	— (5)		
22° bevel	1.27	1.31	0.11 (13)	1.16 (3)	
22° bevel	2.10	1.93	— (3)		
45° bevel	1.47	1.70	0.28 (6)		

From this preliminary study, despite the small sample size, it can be tentatively concluded that under the conditions of reasonable tissue homogeneity and needle sharpness, the diameter of a needle and the geometry of its tip are good predictors of both geometry and size of mode-I cracks. It can also be concluded that depth has little effect on crack size.

From our small sample set, we can predict a branch crack to be roughly 1.5 times larger than needle diameter that caused it. For sharp bevel tips, the crack length is about the same as the diameter of the needle, but for less sharp, short bevel, biopsy needles, as might be expected, the crack size is larger than the needle diameter which accounts for a larger insertion force.

4 Fracture Mechanics Insertion Model

When a sharp object penetrates a tissue, a crack propagates at its tip. An energy balance between two consecutive states assumes the following.

1. Elastic fracture. The deformation is plastic only at the vicinity of the crack.
2. Quasi-static process. Kinetic energy can be neglected, the velocity is low enough for the system to be in equilibrium at all times.
3. Sharp needle interaction. The tip is always in contact with the crack.
4. Constant crack width.

Combining the results in [2,8,14], the energy balance equation gives

$$F\,\mathrm{d}u + \mathrm{d}U_\mathrm{i} = J_{\mathrm{IC}}\,\mathrm{d}A + \mathrm{d}\Delta + \mathrm{d}\Gamma + P\,\mathrm{d}u. \tag{1}$$

In this expression

- $F\,\mathrm{d}u$ is the work done by the needle insertion force F for an increment $\mathrm{d}u$,
- U_i is the internal strain energy stored in the material before indentation,
- $J_{\mathrm{IC}}\,\mathrm{d}A$ is the irreversible work of fracture in mode-I and J_{IC} is the critical fracture toughness (this lumps the different microscopic irreversibilities observed at the tip into a general constant that is material and mode specific).
- $\mathrm{d}A$ is the increment in crack area; for a line mode-I crack of width a, $\mathrm{d}A = a\,\mathrm{d}\,u'$ where $\mathrm{d}u'$ is the crack depth, and $\mathrm{d}\Delta$ is the change in the stored internal recoverable strain energy potential. For instance, during scalpel cutting, the two created surfaces are free and the only strain energy stored is due to the deformation of the tissue in the direction of the cut. On the other hand, during the needle piercing phase, the surface created is also under transverse compression; with a plane-strain assumption, used in Shergold's model [18], the strain energy is stored proportionally to the insertion depth and vertical deformation subsequent to the initial deformation is neglected.
- $P\,\mathrm{d}u$ is the work done by the friction force P along the shaft of the needle. Its magnitude depends on the relative velocity between the needle and the tissue. In a quasi-static insertion, with the constant increase in contact area between the needle and the tissue, P is proportional to the insertion depth.
- $\mathrm{d}\Gamma$ represents the work absorbed in plastic flow. This term is negligeable and is ignored [18]; we assume that once the needle is extracted the body takes back its original geometry and size.

We now apply (1) to the different phases of an insertion. Referring to Fig. 5:

Phase 1 deformation: The needle work creates a deformation δ_i. Assuming that the tissue is initially at rest (1), becomes

$$F\mathrm{d}u = \mathrm{d}\Delta. \tag{2}$$

This force may always be represented as polynomial function $F(\delta)$. This interaction reduces to a contact mechanics problem between a tool and a substrate and for which numerical and analytical solution exists.

Fig. 5. Definition of displacement quantities

Phase 1.5 rupture: When rupture occurs a crack propagates. The maximal deformation δ_i relaxes to δ_c (Fig. 5). During this short instant, the force exhibits a sharp decrease in magnitude and a crack of depth $\delta_i - \delta_c$ is formed. Heverly et al. showed that the force required to initiate cutting reduces with increasing needle velocity, up to a critical speed above which, the rate independent cutting force of the underlying tissue becomes the limiting factor [9]. At this critical rate the tissue deformation is minimal. The limiting velocity is one order of magnitude larger than the data obtained from [13].

Phase 2 penetration: The true crack depth is $u' = u - \delta_c$. Since the tool displacement is equal to the crack propagation, the energy equation becomes

$$F du = J_{\mathrm{IC}}\, dA + d\Delta + P du, \tag{3}$$

where $d\Delta$, the work per unit length to wedge open the crack, could be calculated using deformation models, by considering an expanding wedge of radius, r, from $r = 0$ to R [18]. In Shergold and Fleck's model the crack width is determined as a function of R, J_{IC} and the tissue properties, shear modulus μ and strain hardening α. Assumptions include plane strain, isotropic, non-viscous and incompressible material. Since their model ignores friction, the predicted crack length is probably overestimated.

5   Fracture Toughness Determination

To determine the work of fracture, $F_c\, da$, where F_c is the cutting force, an approach similar to that in [14,16] is employed. A cutting-pass penetration of the needle is first carried out. During insertion, the friction, wedging and fracture work, all contribute to the force F. The energy balance remains that of (3). A second penetration is performed at the same location. All phenomena are nearly identical except for the energy required to create the crack, which vanishes. This gives a smaller needle force F' and (3) reduces to

$$F' du = d\Delta + P\, du. \tag{4}$$

Subtracting (3) from (4) and noting that $du' = du$ we obtain an explicit relationship for the fracture toughness

$$(F - F')\, du = J_{\mathrm{IC}}\, a\, du. \tag{5}$$

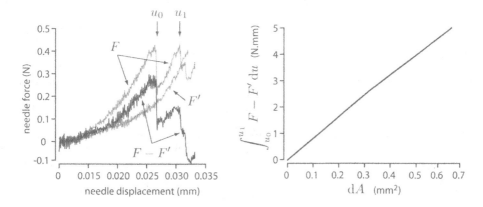

Fig. 6. Representative example of the determination of J_{IC} using integration

Evaluating $\int_{u_0}^{u_1}(F - F')\,du$ as a function of $dA = a\,du$ where u_0 is the onset of the steady penetration phase and u_i is when it terminates. The slope of the estimated curve is the fracture toughness J_{IC} of the pierced tissue. See Fig. 6.

6   Results

In Maurin et al. [13], an 18 gauge, 1.27 mm diameter, 22° biopsy needle was roboticaly inserted in vivo into a pig liver at different slow velocities with a pause of 8 s between the insertion and the extraction phases. During insertion, forces and displacements were recorded.

We applied the model described in the previous section to their data. Based on our in vitro experiments of Section 3.1, the crack length was estimated to be equal to the needle diameter. This assumption is realistic since the difference in tissue condition is expected to marginally affect the crack size.

For each of the four measurements, labeled as in the original reference, the integral $\int_{u_0}^{u_1} F_n - F_n'\,du$ was obtained using the trapezoidal approximation and plotted against the crack area expansion. The slopes obtained are collected in Table 2.

Table 2. Fracture toughness for different measurements

Measurements	$J_{IC}(\text{J·m}^{-2})$	Velocity(mm·s^{-1})
3b	75.8	15
4b	95.7	15
5b	91.8	6
6b	185.6	19

7   Discussion

Measurements 3b, 4b and 5b fall within a narrow range of values while measurement 6b gave a higher value. Such disparity cannot be attributed to the change in velocity but rather to inhomogeneity of the organ. The number of measurements is not statistically significant, but the range of fracture toughness is reasonable 76 to 185 J·m^{-2}. Additional tests would likely confirm these results. By comparison, Chanthasopeephan et al. found for a dead pig liver, using scalpel cutting, a fracture toughness ranging from 187 to 225 J·m^{-2} [6]. This discrepancy is easily explained. Scalpel-liver interactions differ from needle insertion. Differences may also exist in the sharpness of the tool; a sharper tool produces less blunting [14]. The test conditions were also different. Maurin et al.'s experiments were done in vivo with both breathing and non-breathing animals.

The accuracy of ours results is relative to our assumptions and to method employed. The crack width could only be estimated. A 10% error in the crack size estimate would result in a 10% error in fracture toughness. Plastic deformation was also neglected but this assumption is reasonable. When a free-pass was run, it was assumed that the needle experienced the same friction forces and caused the same strain energy storage as a piercing-cut. Data from multiple passes at a same location should be collected to investigate this effect further.

In future experiments, tissue deformation and friction should be modeled, and in combination with fracture mechanics, analytical needle tip forces should be compared to experimental results. In addition, other experimental conditions should be controlled and varied. For example, the introduction of kinetic energy to account for non-steady insertions or taking into account the viscosity of the tissue would certainly improve the model.

8   Conclusion

Needle insertion was modeled as controlled crack propagation. We first confirmed that a planar crack is created when a biopsy needle is inserted into liver and that a star-crack results from using a Franseen tip. By using an energy approach, the fracture toughness of porcine liver was estimated. Our technique can be applied for other needle geometries and tissues. A significant variable is the crack size which can reflect the tip geometry and can be set to simulate a variety of needles. Future work should combine these results with FEM deformation models.

Acknowledgments

The authors wish to thank Benjamin Maurin and Laurent Barbé for sharing their data as well as Thomas Steffen and Janet Moir for help with imaging the samples. This article is based on "Estimation of the penetration force of needle insertion into soft tissue" submitted in partial fulfillment of course "Haptics" taught by the second author at McGill University. The second author would like to acknowledge the support of NSERC, the Natural Sciences and Engineering Research Council of Canada for a Discovery grant.

References

1. Abolhassani, N., Patel, R., Moallem, M.: Needle Insertion Into Soft Tissue: a Survey. Med. Eng. & Phy. 29, 413–431 (2007)
2. Atkins, A.G., Mai, Y.-W.: Elastic and Plastic Fracture, Metals, Polymers, Ceramics, Composites, Biological Materials. Ellis Horwood Limited (1985)
3. Barbé, L., Bayle, B., Mathelin, M., Gangi, A.: Needle Insertions modeling: Identifiability and limitations. Biomed. Sig. Process. Contr. 2, 191–198 (2007)
4. Barbé, L., Bayle, B., Mathelin, M., Gangi, A.: In Vivo Model Estimation and Haptic Characterization of Needle Insertions. Int. J. Robotics Research 26, 1283–1300 (2007)
5. Brett, P.N., Parker, T.J., Harrisson, A.J., Thomas, T.A.: Simulation of Resistance Forces Acting on Surgical Needles. Proceedings of the Institution of Mechanical Engineering 211, 335–345 (1997)
6. Chanthasopeephan, T., Desai, J.P., Lau, A.C.W.: Determining Fracture Characteristics in Scalpel Cutting of Soft Tissue. In: IEEE/RAS-EMBS Conference on Biomedical Robotics and Biomechatronics, pp. 899–904 (2006)
7. Dimaio, S.P., Salcudean, S.E.: Needle Insertion Modeling and Simulation. IEEE T. on Robotics and Automation 19, 191–198 (2003)
8. Doran, C.F., McCormack, B.A.O., Macey, A.: A Simplified Model to Determine the Contribution of Strain Energy in the Failure Process of Thin Biological Membrane During Cutting. Strain 40, 173–179 (2004)
9. Heverly, M., Dupont, P.: Trajectory Optimization for Dynamic Needle Insertion. In: IEEE Int. Conf. on Robotics and Automation, pp. 1646–1651 (2005)
10. Hing, J.T., Brooks, A.D., Desai, J.P.: A biplanar ßuoroscopic approach for the measurement, modeling, and simulation of needle and soft-tissue interaction. Medical Image Analysis 11, 62–78 (2007)
11. Mahvash, M., Hayward, V.: Haptic Rendering of Cutting: a Fracture Mechanics Approach. Haptics-e 2 (2001)
12. Maurel, W.: 3D Modeling of the Human Upper Limb Including the Biomechanics of Joints, Muscles and Soft Tissues. PhD thesis, Laboratoire d'Infographie, Ecole Polytechnique Federale de Lausanne (1999)
13. Maurin, B., Barbe, L., Bayle, B., Zanne, P., Gangloff, J., Mathelin, M., Gangi, A., Soler, L., Forgione, A.: Vivo Study of Forces during Needle Insertions. In: Perspective in Image-guided Surgery: Proceedings of the Scientific Workshop on Medical Robotics, Navigation, and Visualization (MRNV 2004), pp. 415–422 (2004)
14. McCarthy, C.T., Hussey, M., Gilchrist, M.D.: On the Sharpness of Straight Edge Blades in Cuting Soft Solids: Part I - Identation Experiments. Eng. Fract. Mech. 74, 2205–2224 (2007)
15. Okamura, A.M., Simone, C., O'leary, M.D.: Force Modeling for Needle Insertion into Soft Tissue. IEEE T. on Biomedical Engineering 51, 1707–1716 (2004)
16. Pereira, B.P., Lucas, P.W., Teoh, S.-H.: Ranking the Fracture toughness of Thins Mammalian soft Tissues Using the Scissors Cutting Test. J. Biomechanics 30, 91–94 (1997)
17. Shergold, O.A., Fleck, N.A.: Mechanisms of Deep Penetration of Soft Solids, With Application to the Injection and Wounding of Skin. Procedings Royal Society London A 460, 3037–3058 (2004)
18. Shergold, O.A., Fleck, N.A.: Experimental Investigation Into the Deep Penetration of Soft Solids by Sharp and Blunt Punches With Application to the Piercing of Skin. J. Biomechanical Eng. 127, 838–848 (2005)

Towards a Framework for Assessing Deformable Models in Medical Simulation

Maud Marchal, Jérémie Allard, Christian Duriez, and Stéphane Cotin

INRIA Project-team Alcove
IRCICA, 50 avenue Halley
59650 VILLENEUVE D'ASCQ
France

Abstract. Computational techniques for the analysis of mechanical problems have recently moved from traditional engineering disciplines to biomedical simulations. Thus, the number of complex models describing the mechanical behavior of medical environments have increased these last years. While the development of advanced computational tools has led to interesting modeling algorithms, the relevances of these models are often criticized due to incomplete model verification and validation. The objective of this paper is to propose a framework and a methodology for assessing deformable models. This proposal aims at providing tools for testing the behavior of new modeling algorithms proposed in the context of medical simulation. Initial validation results comparing different modeling methods are reported as a first step towards a more complete validation framework and methodology.

1 Introduction

Accurate and interactive simulations of medical environment offer new alternatives and potential helpful tools for the realization of the physician gestures. Thus, deformable models can provide information on the global behavior of soft biological materials, even for locations where it may be difficult to obtain experimental data. In addition, ongoing improvements of computational power make it possible to use more complex models and produce more realistic representations of medical environment. While these motivations have been a driving force for the rapid growth of deformable models, they have also triggered the development of the field of interactive medical simulation. However, in both contexts, a certain level of validation must be established before physicians can use such simulations, whether it is for planning a complex procedure or for learning basic surgical skills. The overall objective of a validation process is to guarantee that: (i) the numerical approximation of the mathematical equations chosen for governing the model is acceptable and (ii) the model provides an accurate representation of the physical behavior of the problem of interest within a given computation time. Both assumptions need to be verified within an assessment of error in the model predictions and their achievement relies on a combination of methodologies and experimental data.

F. Bello, E. Edwards (Eds.): ISBMS 2008, LNCS 5104, pp. 176–184, 2008.

A review on verification, validation and sensitivity studies has recently been proposed in the context of computational biomechanics [1]. In their paper, the authors present the concepts of verification and validation of biomechanical models and introduce a guide to realize such studies. In the context of medical simulation, only some papers have namely proposed solutions to test, compare and quantify the results of different modeling methods, in particular deformable models for soft biological material simulations. Alterovitz et al. [2] have suggested accuracy metrics and benchmarks for comparing different algorithms based on the Finite Element Method (FEM). Validations procedures for discrete approaches have also been introduced [3,4,5]. However these studies are mainly focused on the identification of parameter sets that optimize the accuracy of the discrete models. Real data can also been used as reference models and experiment results have already been presented in the context of medical simulation. Among them, the Truth Cube experiment [6] or experiments on cylinders [7] offer quantitative results, allowing the comparisons of modeling methods with real three-dimensional data. A comparison of FEM simulations with medical images have also been proposed in [8].

All these papers aim at providing reference solutions for either verifying the numerical behavior of models with analytical solutions or validating them against real data. However, the proposed experiments are rarely shared and a methodology based on a combination of a protocol and associated measurements has not been defined yet.

In this paper, a methodology and a framework for assessing deformable models are proposed. The methodology, introduced in Sect. 2, is a combination of analytical models and experimental reference objects that can test the ability of various algorithms to capture a particular deformable behavior. In addition, different metrics are proposed to quantitatively assess the accuracy of algorithms, as well as their computational efficiency. The proposed framework is based on an open source simulation environment where several algorithms are already implemented, thus making it a more consistent basis for comparing algorithms against each other and for validating them against reference models. Its use is illustrated with an example combining analytical solution, real data and different modeling methods in Sect. 3. An initial series of tests illustrate our proposal.

2   Validation Methodology and Framework

2.1   Verification and Validation Protocol

Based on the guide proposed in [1], the protocol for analyzing the performances of a deformable model can be decomposed in two main parts. The first part concerns the verification process of the modeling method. It aims at determining if the model implementation provides a correct description and a solution of the chosen modeling method theory. In this part of the protocol, the benchmarks used to analyze the performances of the model are mainly analytical solutions of well-known problems. Such comparisons have already been proposed in the literature, for example by [2]. In a second stage called validation, the ability of the already verified model to bring a correct simulation of a real world object has to be guaranteed. In this validation part, computational predictions are compared to experimental data as a gold standard.

For both parts of the verification and validation protocol, different types of errors can be identified. The first type concerns numerical errors introduced by computational solving of intractable mathematical equations, among those discretization or convergence errors are very common. This type of error is mainly identified trough the verification process. The second type of error can be called modeling error and is related to assumptions and approximations in the mathematical representation of the physical problem of interest. Such errors mainly come from geometry representation, boundary condition specifications, material properties or the choice of the governing constitutive equations. They can mainly be measured through the validation process.

2.2   Measurements and Metrics

In medical simulators, two types of objectives can be differentiated. A simulator can be dedicated either for a learning task or for the planning of a medical procedure. To validate a deformable model used for the simulation of a medical environment, different performances criteria have to be defined. In the context of medical simulation, we focus on two specific criteria : computational time and accuracy. These criteria allow the evaluation of both the interactivity and the precision offered by a modeling method.

To measure accuracy performances of the different modeling methods, two different types of metrics are proposed, depending on the type of available reference data. For both types, the error can be an absolute value, taking into account the displacement value or a relative value independent of the displacement of the simulated object.

The first type of error metric is used if reference data contain markers (the mesh of an analytical solution or solid markers inside a phantom for example). The metric proposed in this paper is a relative value called the relative energy norm error. This metric is commonly used in the FEM literature [9] and for algorithm comparisons [2]. Let $\mathbf{u}$ be a vector containing the displacement of each point of a discretization of the reference solution and $\hat{\mathbf{u}}$ be a vector containing the simulated displacements of each point of a model (like nodes on a FEM model for example). The relative energy norm error η is defined as:

$$\eta = \|e\|/\|\mathbf{u}\| \tag{1}$$

where $\|e\|$ is the energy norm for the error between the two displacements $\mathbf{u}$ and $\hat{\mathbf{u}}$: $\|e\| = \sqrt{(\hat{\mathbf{u}} - \mathbf{u})^T (\hat{\mathbf{u}} - \mathbf{u})}$.

If reference data do not contain any marker but give information about their global shape (curve, surface, etc), an other error metric based on a measurement of the distance between the reference and the simulated shapes has to be defined. Research works on image registration can provide good metrics. In this paper, we propose a simple distance as a first step for a metric framework.

The measured distance corresponds to the minimal distance between the simulated points sampled on the surface of the simulated model and the surface of the reference data along the normal. This distance can also be normalized by the simulated displacement for each point. The obtained value is realistic only for small errors between the simulated and the reference displacements. This second metric preferentially gives information on the surface of the simulated objects while the first metric provides measurements on the entire volume.

Computational efficiency is also an essential parameter to consider for the assessment of interactive medical simulations. When dealing with dynamic or kinematic systems, a first measure consists in determining if the algorithm can achieve true real-time computation, i.e. the computation time t_{comp} required for a given time step is less or equal to the time step dt used in the time integration scheme of the algorithm: $t_{comp} \leq dt$. Now, to guarantee interactivity, we also must verify that $dt \leq 1/F_c$ where F_c is a critical frequency (typically 25Hz when only visual feedback is considered, and hundreds of Hertz when haptic feedback is needed). With static or quasi-static equations, the real-time criterion $t_{comp} \leq dt$ is irrelevant as the simulation only consists of a sequence of equilibrium states which are independent of the time sampling. However, the second criterion remains necessary, even if defined differently as: $t_{comp} \leq 1/F_c$. Based on these criteria, the definition of a metric could be a combination of measures of these values, pondered by the simulation objectives. Of course, these criteria and metrics are not the only possible means of evaluating computational efficiency, as many factors influence the overall computation time of soft tissue deformation algorithms. Elements such as the integration scheme, the static or dynamic state of the resolution algorithm and furthermore the computer used to solve the simulations (with the use of a CPU or a GPU based algorithm for example) can lead to variations in computational performances. However, measuring such computational performance can only make sense if it is performed whithin a common framework, to ensure a better impartiality in the measurements, as they are often used comparatively against other algorithms or methods.

2.3   Validation Framework

In this paper, a framework is introduced in order to gather both reference models and metrics for assessing a given deformable model behavior. The chosen framework is an open source simulation environment SOFA where several algorithms are already implemented [10]. A validation environment was added to this framework, allowing to share different reference models (which are either analytical solutions or experimental data) and different solutions from existing modeling methods. The description language proposed in [11] to unify the description of the model and loads applied during a simulation is also used in this framework.

3 Example of a Validation Experiment

3.1 Description

In this paper, an example of comparison protocol is developed in order to illustrate how the metrics and the framework introduced in Sect. 2 can be used. The chosen experiment is an elastic beam under gravity, fixed on one side. This test case is widely known in continuum mechanics and has already been used previously for example by [12]. For this experiment, an analytical solution is available allowing for a verification procedure. Furthermore, real data experiments have been conducted in order to achieve a validation procedure. Doing so, different deformable solutions have been compared to these reference solutions.

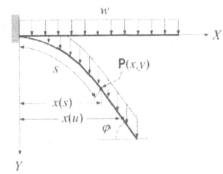

$L(\mathrm{m})$	0.178
$R(\mathrm{m})$	0.019
$\rho(\mathrm{kg.m}^{-3})$	1,070
$E(\mathrm{Pa})$	60,000
ν	0.49
$g(\mathrm{m.s}^{-2})$	9.81

Fig. 1. Description of the beam experiment: definition and values of the parameters

Concerning the beam geometry, the beam cross-section is circular with a radius R and L is the length of the beam. Its density is called ρ, its Young Modulus E and its Poisson ratio ν. The experiment and parameter values are given in Fig. 1.

3.2 Analytical Solution

The Bernoulli-Euler theory of beams provides a formulation of the beam deflection, based on the assumption that a relationship between the bending moment M and the beam curvature κ exists: $\kappa = M/EI$ where I is the moment of inertia of the beam cross-section. For large deflections like in our experiment, the exact expression of the curvature is $\kappa = d\varphi/ds$ where s corresponds to the arc length between the fixed end and a point on the beam and $\varphi(s)$ is the slope of the beam at any point with respect to horizontal (see Fig. 1). If we differentiate the curvature expression with respect to s, we can obtain the differential equation that governs large deflections of the beam:

$$\frac{d^2\varphi}{ds^2} = \frac{1}{EI}\frac{dM}{ds} \tag{2}$$

The bending moment M at a point P of coordinates (x, y) for the deflected beam is given by: $M(s) = \int_s^L w(x(u) - x(s))du$ where $w = \rho g \Pi R^2$ represents the load corresponding to the gravity, uniformly distributed along the entire length. By differentiating this equation once with respect to s, taking into account the relation $cos\varphi = dx/ds$, and substituting it in Equation 2, we obtain the non-linear differential equation that governs the deflections of the beam under the gravity:

$$\frac{d^2\varphi}{ds^2} = -\frac{1}{EI}w(L - s)cos\varphi \tag{3}$$

The numerical solution of this problem was achieved using Maple software (Maple 11.0) and the solutions were compared to the different modeling methods.

3.3 Real Data

Our experimental reference model is a cylindrical beam made of silicone gel. To obtain a material with nearly linear elastic properties, we used a silicon rubber called ECOFLEX (Ecoflex0030).

The estimated Young modulus E is equal to $60,000$Pa and the Poisson ratio has a value of 0.49, as the material is considered as nearly incompressible. The beam was glued on one extremity to an inverted T-shaped vertical support made of plexiglas, and submitted to its own weight. It was then scanned in a helical CT scanner to produce of volume of $512 \times 512 \times 113$ voxels, with a voxel size of $0.425 \times 0.425 \times 0.62$mm^3. The shape of the deformed beam, illustrated in Fig. 2, was reconstructed using a Marching Cube algorithm to produce the reference surface and smoothed to remove noise artefacts.

3.4 Modeling Methods

Our framework allows to gather the analyical solution, the experimental data and different existing modeling methods. In this paper, five different algorithms were compared: (a) a linear FEM algorithm with a tetrahedral mesh, (b) a co-rotational FEM algorithm also with a tetrahedral mesh [13], (c) a co-rotational FEM algorithm with an hexahedral mesh, (d) an algorithm based on 6 Degrees of Freedom Beams [14], (e) a mass-spring network.

3.5 Results

The resulting simulations of different modeling methods available in the SOFA framework have been compared against both an analytical solution and experimental data. At the exception of the mass-spring model for which stiffness coefficients were adjusted to obtain the best behavior, the physical parameters used in all simulations correspond exactly to those measured on the experimental data. Concerning the comparisons with the analytical solution, the relative energy norm error can be used as the different meshes of the models and the analytical solution have the same number of nodes. The results are given in Fig. 3 and Table 1 for the position of tip of the beam. As for comparisons with

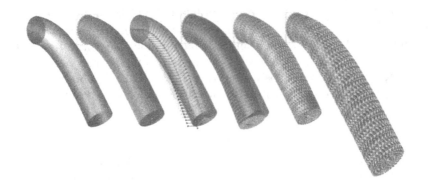

Fig. 2. Simulation results for the different modeling methods. From left to right: the experimental solution, the mass-spring network, the 6-DOF beam, the FEM Hexahedra, the FEM corotational Tetrahedra and the linear FEM Tetrahedra solutions.

Table 1. Comparisons between the different simulations: (a) the linear FEM tetrahedra algorithm, (b) the FEM Tetrahedra co-rotational algorithm, (c) the FEM hexahedra co-rotational algorithm, (d) the 6-DOF Beam algorithm and (e) the mass-spring network. The relative energy norm error is expressed in percentage and relies on the comparisons with the analytical solution. The surface error corresponds to the comparisons with the real data : the absolute value (without normalization) and the percentage are provided (the mean value and the standard deviation).

	Relative Energy Norm Error (%)	Relative Surface Error		
		Mean (mm)	Mean (%)	Standard Deviation
(a)	71.41	18.60	20.34	32.31
(b)	1.15	0.63	4.95	15.44
(c)	11.29	2.87	8.41	14.73
(d)	1.80	4.32	8.09	7.75
(e)	1.41	0.75	4.64	10.72

the experimental data, since no physical markers were used to track volumetric displacements, a measure of the relative surface error was used for the results reported in Table 1. Images of the simulated beams for the different algorithms are given in Fig. 2.

These results confirm, through quantitative measurements, important points about soft tissue modeling algorithms. First, if the underlying model is not appropriate, it is impossible to capture the deformation of the reference model, no matter the choice of parameters. This is well illustrated with the case of the linear elastic FEM model which cannot handle large displacements. On the other hand, our examples also show that it is possible to obtain rather good approximations of a given behavior using different methods (mass-spring model, co-rotational FEM, beam model) all within a range of computation times compatible with interactive simulations. We can also see that even an ideal, analytical model will not give the exact same result as an experiment, some of these differences coming from errors on the various measurements done on the experimental model. Our preliminary results also show the need for a variety of reference models, able to characterize various aspects of soft tissues, to clearly determine which algorithm is best for representing a particular behavior (linear elastic, viscoelastic, bi-phasic, porous, etc...). Similarly, it is important to define metrics that are most relevant to which property of an algorithm we want to evaluate. We have proposed an initial set of metrics to assess the accuracy of the models through comparisons with two different types of reference models. Additional metrics could certainly be proposed, in particular to evaluate the computational efficiency of a particular algorithm.

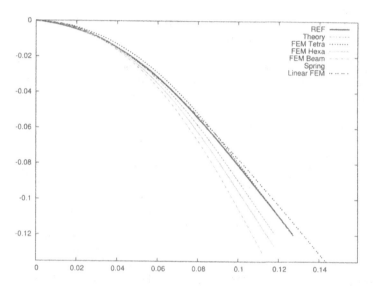

Fig. 3. Central line positions in Cartesian coordinates for each simulated beam, the analytical solution (Theory) and the experimental data (REF)

4   Conclusion

In this paper, we have introduced the basis of a methodology and framework for assessing deformable models in the context of medical simulation. The proposed methodology differs from existing protocols as medical simulators need to be validated both in terms of accuracy and computational efficiency. Although the set of metrics and reference models presented in this paper is limited, we believe they illustrate well the importance to quantitatively assess algorithms used in medical simulation. However, the main novelty of our approach lies in the combination of a unified, open framework where all models could be compared, new metrics defined, algorithms and reference models added. This will eventually enable an unbiased comparison of the performance and accuracy of many different algorithms, to create an Open Benchmark for medical simulation. To this end, we have planned to release the data, models and algorithms used in this study as part of the next SOFA release.

Acknowledgments

This work was supported by the French National Institute for Research in Computer Science and Control (INRIA). We thank J.P. de la Plata Alcalde, Drs J. Dequidt and A. Theetten for their help and the SOFA team for their software help.

References

1. Anderson, A., Ellis, B., Weiss, J.: Verification, validation and sensitivity studies in computational biomechanics. Computer Methods in Biomechanics and Biomedical Engineering 10(3), 171–184 (2007)
2. Alterovitz, R., Goldberg, K.: Comparing algorithms for soft tissue deformation: Accuracy metrics and benchmarks. Technical report, UC Berkeley (2002)
3. Bianchi, G., Solenthaler, B., Szkely, G., Harders, M.: Simultaneous topology and stiffness identification for mass-spring models based on fem reference deformations. In: Barillot, C., Haynor, D.R., Hellier, P. (eds.) MICCAI 2004. LNCS, vol. 3217, pp. 293–301. Springer, Heidelberg (2004)
4. van Gelder, A.: Approximate simulation of elastic membranes by triangulated spring meshes. Journal of Graphics Tools 3(2), 21–41 (1998)
5. Baudet, V., Beuve, M., Jaillet, F., Shariat, B., Zara, F.: Integrating tensile parameters in 3d mass-spring system. In: Proceedings of Surgetica (2007)
6. Kerdok, A., Cotin, S., Ottensmeyer, M., Galea, A., Howe, R., Dawson, S.: Truthcube: Establishing physical standards for real time soft tissue simulation. Medical Image Analysis 7, 283–291 (2003)
7. Leskowsky, R., Cooke, M., Ernst, M., Harders, M.: Using multidimensional scaling to quantify the fidelity of haptic rendering of deformable objects. In: Proceedings of EuroHaptics, pp. 289–295 (2006)
8. Chabanas, M., Payan, Y., Marcaux, C., Swider, P., Boutault, F.: Comparison of linear and non-linear soft tissue models with post-operative ct scan in maxillofacial surgery. In: International Symposium on Medical Simulation, pp. 19–27 (2004)
9. Zienkiewicz, O.C., Taylor, R.L.: The Finite Element Method, 5th edn. Butterworth-Heinemann (2000)
10. Allard, J., Cotin, S., Faure, F., Bensoussan, P.J., Poyer, F., Duriez, C., Delingette, H., Grisoni, L.: Sofa - an open source framework for medical simulation. In: Proceedings of Medecine Meets Virtual Reality, pp. 13–18 (2007)
11. Chabanas, M., Promayon, E.: Physical model language: Towards a unified representation for continuous and discrete models. In: Proceedings of International Symposium on Medical Simulation, pp. 256–266 (2004)
12. Picinbono, G., Delingette, H., Ayache, N.: Non-linear and anisotropic elastic soft tissue models for medical simulation. In: Proceedings of IEEE International Conference on Robotics and Automation, pp. 1370–1375 (2001)
13. Muller, M., Gross, M.: Interactive virtual materials. In: Proceedings of Graphics Interface, pp. 239–246 (2004)
14. Duriez, C., Cotin, S., Lenoir, J., Neumann, P.: New approaches to catheter navigation for interventional radiology simulation. Computer Aided Surgery 11(6), 300–308 (2006)

Image Guidance and Surgery Simulation Using Inverse Nonlinear Finite Element Methods

Philip Pratt

Institute of Biomedical Engineering
Imperial College of Science, Technology and Medicine
London, United Kingdom
p.pratt@imperial.ac.uk

Abstract. Nonlinear finite element methods are described in which cyclic organ motion is implied from 4D scan data. The equations of motion corresponding to an explicit integration of the total Lagrangian formulation are reversed, such that the sequence of node forces which produces known changes in displacement is recovered. The forces are resolved from the global coordinate system into systems local to each element, and at every simulation time step are expressed as weighted sums of edge vectors. In the presence of large deformations and rotations, this facilitates the combination of external forces, such as tool-tissue interactions, and also positional constraints. Applications in the areas of surgery simulation and minimally invasive robotic interventions are proposed, and the methods are illustrated using CT images of a pneumatically-operated beating heart phantom.

1   Introduction

The future of minimally invasive robotic surgery lies not only in the mechanical evolution of better tele-manipulator systems, but also in the development of advanced software tools that facilitate surgical training, patient-specific intraoperative rehearsal, and the seamless integration of preoperative and intraoperative imaging, of various modalities, through augmented reality techniques. The image-constrained biomechanical modelling (ICBM) approach is a key technology which promises to realise the goal of allowing surgeons to alternate between full surgical simulation, endoscopic views enhanced in real-time through the same simulation constrained by imaging data, and completion of the intervention itself. To that end, this paper extends previous work [1] by showing how intrinsic cyclic tissue motion can be inferred from 4D scan data and combined with externally induced motion and other constraints using a *nonlinear* finite element model. Taking known changes in node displacements over time, the finite element model inverted such that the sequence of node forces responsible for the motion can be recovered. These forces are then resolved from the global coordinate system into systems local to each element, thereby expressing them in terms of local geometry. That way, in the presence of large deformations and rotations, external forces and other constraints can be combined when forward simulation is performed.

F. Bello, E. Edwards (Eds.): ISBMS 2008, LNCS 5104, pp. 185–190, 2008.

2   Methods

The equations of motion for a mesh comprising N nodes are expressed in terms of the displacements from the initial configuration, i.e. $\mathbf{U} = [u^0\, u^1\, \ldots\, u^{3N-1}]^\mathsf{T}$, and following the notation of Bathe [2], are written in semi-discrete form as

$$\mathbf{M}\,{}^t\ddot{\mathbf{U}} + \mathbf{C}\,{}^t\dot{\mathbf{U}} + {}^t_0\mathbf{F} = {}^t\mathbf{R} \tag{1}$$

where $\mathbf{U}$, ${}^t\dot{\mathbf{U}}$ and ${}^t\ddot{\mathbf{U}}$ are the displacement, velocity and acceleration vectors, respectively, $\mathbf{M}$ is the mass matrix, $\mathbf{C}$ is the damping matrix, ${}^t_0\mathbf{F}$ is the vector of nodal reaction forces equivalent to the element stresses, and ${}^t\mathbf{R}$ is the vector of externally applied, time-varying forces. The damping matrix is assumed to be proportional to the mass matrix, i.e. $\mathbf{C} = \alpha\mathbf{M}$, where α is the damping coefficient. The mass matrix is assumed to be constant, and is diagonalised to facilitate explicit integration.

2.1   Total Lagrangian Formulation

In the total Lagrangian formulation of the finite element method [3], quantities are expressed in terms of the reference configuration. Considering an individual element i, the nodal reaction forces are computed as an integral over the element volume, as follows

$$ {}^t_0\mathbf{F}^{(i)} = \int_{{}^0V^{(i)}} {}^t_0\mathbf{B}_L^\mathsf{T}\, {}^t_0\hat{\mathbf{S}}\, d^0V \tag{2}$$

where ${}^t_0\mathbf{B}_L$ is the full strain-displacement matrix and ${}^t_0\hat{\mathbf{S}}$ is the vector of 2nd Piola-Kirchoff stresses. The latter depend on the element deformation and the choice of material constitutive law. For an assemblage of elements, the nodal reaction forces are accumulated in accordance with the mesh's element-node relationships.

2.2   Explicit Central Difference Integration

Over a single cycle, the fully discretised equations of motion take the form shown in (3). Note that the nodal reaction forces must be calculated at every time step. At the expense of some restriction on time step magnitude, the explicit scheme avoids the iterative solution of the displacements at the next time step, which would otherwise be extremely computationally expensive. Over multiple cycles, this previous and next displacement vectors wrap around in a straightforward manner.

$$ {}^{next(t)}\mathbf{U} \approx \frac{2}{2 + \alpha\Delta t}\left[\frac{\Delta t^2}{\mathbf{M}}({}^t\mathbf{R} - {}^t_0\mathbf{F}) + 2\,{}^t\mathbf{U} - (\frac{\alpha\Delta t}{2} - 1)^{prev(t)}\mathbf{U}\right] \tag{3}$$

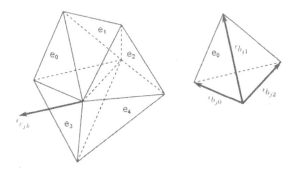

Fig. 1. Force resolution using local geometry

2.3   Recovering Forces from Displacements

By inverting equation (3), one can write the discretised applied force in terms of the displacement, velocity and acceleration vectors, the nodal reaction forces, and other known quantities. By construction, if one were then to solve the equations of motion and apply these forces at the appropriate times, one would recover the original cyclic motion exactly and indefinitely. Note that the recovered forces are expressed in the global coordinate system. The following two-stage calculation is performed at each simulation time step in the cycle.

For each integration point in each element:

- Calculate deformation gradient tensor ${}^t_0\mathbf{X}$
- Calculate strain-displacement matrix ${}^t_0\mathbf{B}_L = {}^t_0\mathbf{B}_{L0}\, {}^t_0\mathbf{X}^\top$
- Calculate 2nd Piola-Kirchoff stress vector ${}^t_0\hat{\mathbf{S}}$
- Accumulate element nodal reaction forces ${}^t_0\mathbf{F}^{(i)}$ to give node totals ${}^t_0\mathbf{F}$

For each node:

- Invert displacement update step to obtain recovered node forces:

$$ {}^t\mathbf{R} \approx \frac{\mathbf{M}}{\Delta t^2} \left[(1 + \frac{\alpha \Delta t}{2})^{next(t)}\mathbf{U} - 2\, {}^t\mathbf{U} + (1 - \frac{\alpha \Delta t}{2})^{prev(t)}\mathbf{U} \right] + {}^t_0\mathbf{F} \qquad (4) $$

2.4   Local Force Resolution

In order to combine recovered and external forces, the former must be expressed not in the global coordinate system, but for each node in terms of its local surrounding geometry. With the introduction of external forces, the geometry may deviate from the original cyclic motion through potentially large-scale deformations and rotations. By resolving recovered forces locally, they are made to act in the appropriate direction in conjunction with externally induced motion.

   The recovered force acting on a particular node in the mesh is assumed to originate from the elements which contain that node. Indeed, an approximation is made whereby the force receives an equal contribution from each such element. For a given element at each point in time, the edge vectors from the node in question to the other three nodes in that element define a local basis in terms of which that element's fraction of the node force can be expressed. This amounts to equating the force to a weighted sum of those edge vectors and solving for the weights. Subsequently, the weights are further computed over the node's other parent elements, ultimately building a set of weights that links all the recovered forces locally to the geometry of the entire mesh.

Labelled with coordinate indices j, figure 1 (left) depicts at time t a typical node k with its recovered force ${}^t r_{jk}$, and the five surrounding elements $e_0, ..., e_4$ to which it belongs. In general, a node will be common to M_k elements. Figure 1 (right) illustrates the first element e_0, and the three edge vectors ${}^t b_{j0}$, ${}^t b_{j1}$ and ${}^t b_{j2}$ which, with the node itself, define the geometry of the element at that instant. In order to express the required fraction of the force in terms of local mesh geometry, its components in the global coordinate system are equated to weighted combinations of these edge vectors, as shown in (5).

The weights are determined by direct inversion of the left-hand-side matrix. This process is repeated for the element's other three nodes, and subsequently over all nodes in the mesh. In order to ensure that the magnitudes of the recovered forces, expressed as functions of current element edge vectors, remain within reasonable bounds and do not give rise to simulation instability, recovered forces are normalised at each time step to have the same magnitude as those implied from the original motion where no externally applied forces are present.

$$
\begin{bmatrix} {}^t b_{00} & {}^t b_{01} & {}^t b_{02} \\ {}^t b_{10} & {}^t b_{11} & {}^t b_{12} \\ {}^t b_{20} & {}^t b_{21} & {}^t b_{22} \end{bmatrix} \begin{bmatrix} {}^t w_0 \\ {}^t w_1 \\ {}^t w_2 \end{bmatrix} = \frac{1}{M_k} \begin{bmatrix} {}^t r_{0k} \\ {}^t r_{1k} \\ {}^t r_{2k} \end{bmatrix} \tag{5}
$$

3   Results

The force recovery and resolution techniques are illustrated using data taken from scans of a beating heart phantom, using an isotropic, hyperelastic neo-Hookean tissue model. The Chamberlain Group CABG phantom was scanned at 54 bpm with a Philips 64-slice CT scanner, producing 10 uniformly-spaced phases. The first of these was manually segmented and converted into a tetrahedral mesh using the SimBio-Vgrid [4] mesh generator. The Image Registration Toolkit [5,6] was used to create a sequence of 3D tensor product cubic B-spline deformations, mapping the initial mesh onto each phase in turn. Cyclic cubic B-splines, defined using 6 uniformly spaced knots, were then used to interpolate mesh node positions over time.

The following constants were used in the simulation: Young's modulus $E = 3.0\text{E} + 03\,\text{Pa}$; Poisson's ratio $\nu = 0.45$; material density $\rho = 1.0\text{E} + 03\,\text{kg/m}^3$; and mass damping coefficient $\alpha = 7.5\text{E}+01$. Thus, $\lambda \approx 9.31\text{E} + 03\,\text{Pa}$ and $\mu \approx 1.03\text{E} + 03\,\text{Pa}$. The equations of motion were integrated using a time step of $\Delta t = 0.001$ seconds.

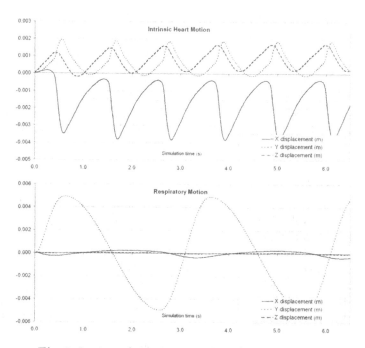

Fig. 2. Intrinsic (upper) and respiratory (lower) motion

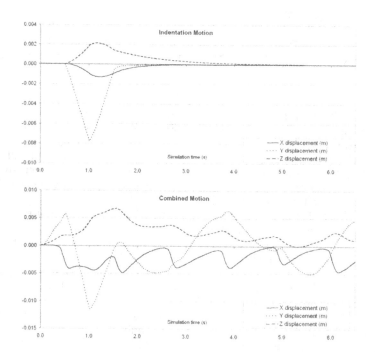

Fig. 3. Indentation (upper) and composite (lower) motion

Figure 2 (upper) shows the motion of a typical surface node as a result of applying recovered forces alone. Unlike the heart phantom itself, several nodes in the base of the mesh are deliberately anchored, and hence it takes a short period of time to converge to an exactly repeatable motion. Figure 2 (lower) and figure 3 (upper) show the individual effects of displacement constraints simulating respiratory motion, and externally applied forces simulating tool-tissue interaction, respectively. Finally, figure 3 (lower) illustrates how all three motions are combined by the model. The underlying beating motion is apparent throughout the simulation.

4   Conclusion

This paper describes a novel technique for constructing nonlinear finite element simulations with cyclical motion recovered from 4D tomographic scan data, whereby external forces and positional constraints can be combined to produce realistic composite behaviour. The technique has immediate applications in the field of patient-specific surgery simulation, and will also form the basis of an image-constrained biomechanical modelling approach to intraoperative image guidance. Future work includes validation of the technique with real patient data, an efficient GPU implementation, anisotropic weighting during the force recovery procedure, and the introduction of weight interpolation and scaling such that the frequency and intensity of the cyclic motion can be modified.

Acknowledgements

The Image Registration Toolkit was used under licence from IXICO Ltd.. The author would like to thank staff at St. Mary's and the Royal Brompton hospitals.

References

1. Pratt, P., Bello, F., Edwards, E., Rueckert, D.: Interactive finite element simulation of the beating heart for image-guided robotic cardiac surgery. In: Medicine Meets Virtual Reality, vol. 16, pp. 378–383. IOS Press, Amsterdam (2008)
2. Bathe, K.J.: Finite element procedures, 1st edn. Prentice-Hall, Englewood Cliffs (2007)
3. Miller, K., Joldes, G., Lance, D., Wittek, A.: Total lagrangian explicit dynamics finite element algorithm for computing soft tissue deformation. Communications in Numerical Methods in Engineering 23, 121–134 (2007)
4. Berti, G.: Image-based unstructured 3D mesh generation for medical applications. In: European Congress on Computational Methods in Applied Sciences and Engineering - ECCOMAS 2004, University of Jyväskylä, Department of Mathematical Information Technology (2004)
5. Rueckert, D., Sonoda, L.I., Hayes, C., Hill, D.L.G., et al.: Non-rigid registration using free-form deformations: Application to breast MR images. IEEE Transactions on Medical Imaging 18(8), 712–721 (1999)
6. Schnabel, J.A., Rueckert, D., Quist, M., Blackall, J.M., et al.: A generic framework for non-rigid registration based on non-uniform multi-level free-form deformations. In: Niessen, W.J., Viergever, M.A. (eds.) MICCAI 2001. LNCS, vol. 2208, pp. 573–581. Springer, Heidelberg (2001)

Laparoscopic Tool Tracking Method for Augmented Reality Surgical Applications

Alicia M Cano, Francisco Gayá, Pablo Lamata, Patricia Sánchez-González, and Enrique J. Gómez

Grupo de Bioingeniería y Telemedicina (GBT),
ETSIT, Universidad Politécnica de Madrid
C/ Ciudad Universitaria s/n, 28040 Madrid, Spain
{acano, fgaya, lamata, psanchez, egomez}@gbt.tfo.upm.es
http://www.gbt.tfo.upm.es

Abstract. Vision-based tracking of laparoscopic tools offers new possibilities for improving surgical training and for developing new augmented reality surgical applications. We present an original method to determine not only the tip position, but also the orientation of a laparoscopic tool respect to the camera coordinate frame. A simple mathematical formulation shows how segmented tool edges and camera field of view define the tool 3D orientation. Then, 3D position of the tool tip is determined by image 2D coordinates of any known point of the tool and by tool's diameter. Accuracy is evaluated in real image sequences with known ground truth. Results show a positioning error of 9,28 mmRMS, what is explained by inaccuracies in the estimation of tool edges. The main advantage of proposed method is its robustness to occlusions of the tool tip.

Keywords: laparoscopic tool tracking, video-endoscopic analysis, augmented reality tools, surgical training programs.

1 Introduction

Laparoscopic surgery involves an important challenge for surgeons in getting used to the reduced workspace and the limited sensory interaction [1]. This technique offers a different interaction paradigm to the traditional open approach, based on the distance manipulation of specialized surgical instruments and an indirect visualization of the surgical scene captured by an endoscope. This makes the learning and developing of the technique a difficult task.

3D localization of surgical instruments opens the possibility of multiple applications for overcoming some of the current limitations of laparoscopy. Surgical training programs lack of standardized and assessed methods with objective metrics to guarantee the complete instruction of surgeons [2]. The analysis of trajectory and movements of tools offers objectives parameters, suitable for defining objective metrics [3],[4]. In particular, new augmented reality devices for training can exploit the benefits of this tracking to provide trainees with constructive and effective feedback about their performance [5].

F. Bello, E. Edwards (Eds.): ISBMS 2008, LNCS 5104, pp. 191–196, 2008.

On the other hand, the last advances on augmented reality applied in surgery provide useful information for image-guidance for intra-operative procedures in minimally invasive surgery. In these techniques, the knowledge of the tools 3D position constitutes an important step to guide and advice surgeon about the proximity of delicate areas [5],[6].

In this work we address the analysis of laparoscopic video sequences as the main source of information to get 3D information of the surgical scene. We aim to achieve 3D tracking of laparoscopic tools using the 2D information of the image. This is an interesting alternative to other means to localize instruments based on the use of position sensor on the tools (optical, electromagnetic or mechanical), which can be very bulky [7].

In the scientific literature, a previous approach [8] uses a LED placed in the tool tip that projects laser dots onto the surface of the organs. The optical markers are then detected in the endoscope image on the surface to allow localizing the instrument with respect to the scene. Other more recent works have tackled the problem without extra markers. Tonet et al. [9] proposed a heuristic approach to estimate the depth of the instrument by means of the knowledge of the tool width and the orientation of its edges. Authors have previously formalised two methods for assessing the 3D position of the tools' tips [10]. The first uses the vanishing point of the tool in the image, and the second one the tool width at the tip in the image.

Video-based localization of laparoscopic tools requires two main steps: extraction of relevant 2D information from the image, and calculation of the 3D coordinates of the tool's tip. This paper focuses on the second, and contributes with a new method that calculates not only tool's position, but also orientation. The method is also robust to the occlusion of the tool tip.

2   Materials and Methods

The 2D relevant information is composed by the tool edges and tip, which are segmented from laparoscopic images. Tool edges detection strategy is based on the temporal continuity between consecutives frames and edge detector operator (Sobel). Colour analysis of the laparoscopic tool allows us to identify a point of the tool tip.

The second stage of the proposed method is explained in detail in the next section.

2.1   3D Localization Method for Laparoscopic Tool

Segmented tool edges and camera field of view (FoV) define the tool 3D orientation. Then, 3D position of the tool tip is determined by image 2D coordinates of the tool tip (or any characteristic point of the tool as we explain later) and by tool's physical diameter (5 or 10 mm usually in laparoscopic tools).

Let C be the optical camera centre, P a point of the laparoscopic tool axis, and u_{CP} unitary vector from C to P. Let Π be the image plane, and (Ω_1, Ω_2) the tangential planes to the tool containing C. The intersection of both (Ω_1, Ω_2) planes with Π plane results in E_1 and E_2, the two projective edges of the tool (see fig 1.a), being $(\vec{u}_{E1}, \vec{u}_{E2})$ their unitary vectors.

Let $(\vec{u}_{CE1}, \vec{u}_{CE2})$ be unitary vectors from C to any point of E_1 or E_2 respectively. Each Ω_i plane normal vector, $(\vec{u}_{\Omega 1}, \vec{u}_{\Omega 2})$, is obtained from:

$$\vec{u}_{\Omega i} = \vec{u}_{Ei} \times \vec{u}_{CEi} \ , \ i = 1, 2 \tag{1}$$

because both vectors are included in Ω_i plane.

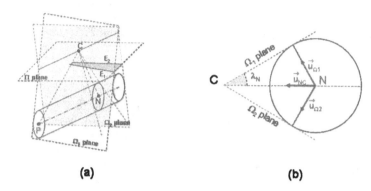

(a)   **(b)**

Fig. 1. (a) Projective model of the laparoscopic tool; C: optical center, P: tip of the laparoscopic tool; N: point of the cylinder axis which projective line is perpendicular to this axis; Ω_1, Ω_2: planes of sight of the tool edges; Π plane: image plane, E1, E2: projective tool edges, (b) Transversal section of the laparoscopic tool at N point

The perpendicular plane to the cylinder's axis that contains C defines a circle section of the cylinder as shown in fig. 1.b. Let N be the centre of this circle, and $2\lambda_N$ the angle between planes Ω_1 and Ω_2.

As shown in fig. 1.b the addition $\vec{u}_{\Omega 1}$ and $\vec{u}_{\Omega 2}$, results in the direction of $\overline{CN}$, and its cross product gives the tools' direction $\vec{u}_{NP}$. Therefore angle λ_N is expressed like:

$$\tan \lambda_N = \frac{\left| \vec{u}_{\Omega 1} + \vec{u}_{\Omega 2} \right|}{\left| \vec{u}_{\Omega 1} - \vec{u}_{\Omega 2} \right|} \tag{2}$$

Equally, it is possible to determine |CN| from the tool radius (r_{CYL}):

$$|CN| = r_{CYL} / \sin \lambda_N \tag{3}$$

Finally, CP is found with:

$$\overline{CP} = \frac{|CN| \cdot \vec{u}_{CP}}{\vec{u}_{CP} \bullet \vec{u}_{CN}} \tag{4}$$

Note that P can be any point of the tool, it only requires to be identified in the image. And as $\vec{u}_{NP}$ is known, any other point of the tool P' (the tip) can be obtained by knowing the physical distance between P and P'.

2.2 Validation Workbench

Laboratory laparoscopic setting has been built to simulate the real surgical geometry and the range of movements of the tool. This scenery allows us to determine the ground truth of the tool positions thanks to the introduction of a background plane (a grid of points) and an inclined board (30°). This board is placed at a known distance from the optical centre point, knowing the position of any point of the plane respect to the camera. This way the 3D trajectory of the tool keeping the tip alongside this board is defined with only 2D information gathered from the segmentation stage.

Two validation sequences are acquired, performing two different movements: one with a constant depth ("Constant depth" sequence) and the second one with a variable depth ("Variable depth" sequence). Each sequence has about 275 frames.

3   Results

The two lab sequences have been analysed with proposed method (we call it Transversal Section based method) and with two previous ones [10], based on the Vanishing Point and on the Apparent Diameter. Accuracy of assessed 3D coordinates is given by comparing them to the Ground Truth known by the 2D position of the tool in the image. Results are reported in table 1 and fig 2.

Table 1. Error characterization of methods for validation sequences. Mean error and standard deviation (SD), in mm.

	Axis	Constant depth		Variable depth	
		Mean	SD	Mean	SD
Vanishing Point	X	2.29	1.81	0.98	7.2
	Y	3.43	11.09	1.39	3.42
	Z	42.16	22.50	33.22	9.97
Apparent Diameter	X	0.71	0.15	0.60	0.76
	Y	1.93	0.86	0.24	3.55
	Z	2.39	2.61	6.32	4.36
Transversal Section	X	0.08	0.08	0.03	1.4
	Y	0.35	1.25	1.54	3.66
	Z	2.03	2.09	8.65	3.55

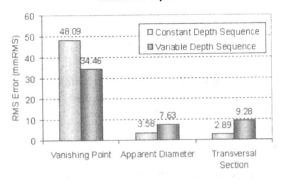

Fig. 2. Root Mean Square Error (mmRMS) in Z coordinates for three methods

4 Discussion

A new method for video-based tracking of laparoscopic tool has been proposed. It is robust to tip occlusions providing that a visible characteristic point of the tool is visible in the image. It determines not only the tip position, but also the orientation of a laparoscopic tool respect to the camera coordinate frame.

Validation sequences in a controlled laboratory setting were taken to enable the estimation of the ground truth of tool's positions. This process might be subjected to biases due to errors in calibration or in the image segmentation of the tool. Nevertheless, we think this practical solution is valid to present the relative performance of the three methods and a first approximation of its tracking accuracy. Our future work will consider the acquisition of a reliable ground truth and the use of barrel distortion correction in laparoscopic sequences to get an accurate estimation of the performance of these video-based tracking methods.

Results have verified the potential of video-based tracking methods, as already concluded in previous works [9],[10]. Proposed method offers a good accuracy (error of 2,89 mmRMS and 9,28 mmRMS for the each of the two sequences, see fig.2). The origin of this error is explained with possible calibration issues.

Measured accuracy of proposed Transversal Section method does not improve what the Apparent diameter method achieves (see fig.2). Nevertheless, the latest requires a visualization without obstacles of the tool tip, limiting its potential use respect to proposed method. Moreover, a combination of both methods could improve the behaviour of each other, since the 3D localization is determined through different features of the tool.

A limiting factor in image-based tracking of tools is the presence of blurred tool's edges caused by its movement. This causes an inaccurate edge detection, and an error in a later 3D pose estimation. A robust edge detection can include temporal constrains, an also geometrical constrains based on the position and projection of the insertion point (trocar point) [11], what is easily found with the 3D orientation information extracted from proposed method.

Current tracking performance is good enough for gesture analysis and an objective evaluation of surgical manoeuvres, where non real time is needed and accuracy around 3 mm. is enough to determine objective metrics. An improvement of the method addressing the points commented before will increase its potential applications in the clinical routine. In particular, this tools tracking appears as one of the keys to development of new augmented reality tools for laparoscopic surgery.

5 Conclusion

This work proposes a new method for video-based tracking of laparoscopic tools that obtains both their position and orientation. It is also robust to occlusions of the tip equipping the tool with an additional landmark.

Results in laboratory environment have validated the proposed method as an alternative to traditional tracking, with important applications such as an objective evaluation of motor skills and the development of augmented reality environments.

Acknowledgments. Authors would like to express their gratitude to Dr. F. Sánchez-Margallo from Minimally Invasive Surgery Centre (Cáceres, Spain), for his valuable contributions to this work.

References

1. Fuchs, K.H.: Minimally Invasive Surgery. Endoscopy 34, 154–159 (2002)
2. Darzi, A., Smith, S., Taffinder, N.: Assessing operative skill: needs to become more objective. BMJ 316, 1917–1918 (1998)
3. Rosen, J., Brown, J.D., Chang, L., Sinanan, M.N., Hannaford, B.: Generalized Approach for Modeling Minimally Invasive Surgery as a Stochastic Process Using a Discrete Markov Model. IEEE Transactions on Biomedical Engineering 53(3), 399–413 (2006)
4. Aggarwal, R., Grantcharov, T., Moorthy, K., Milland, T., Papasavas, P., Dosis, A., Bello, F., Darzi, A.: An evaluation of the feasibility, validity and reliability of laparoscopic skills assessment in the operating room. Ann. Surg. 245(6), 992–999 (2007)
5. Blum, T., Sielhorst, T., Navab, N.: Advanced augmented reality feedback for teaching 3D tool manipulation. New Tech. frontiers in minimally invasive therapies, pp. 223-236 (2007)
6. Fuchs, K.H., Livingston, M.A., Raskar, R., D'nardo, C., Keller, K., State, A., Crawford, J.R., Rademacher, P., Drake, S.H., Meyer, A.A.: Augmented Reality Visualization for laparoscopic surgery. In: Wells, W.M., Colchester, A.C.F., Delp, S.L. (eds.) MICCAI 1998. LNCS, vol. 1496, pp. 934–943. Springer, Heidelberg (1998)
7. Immersion: Digiziting, http://www.immersion.com/digitizer/
8. Krupa, A., Gangloff, J., Doignon, C., de Mathelin, M., Morel, G., Leroy, J., Soler, L., Marescaux, J.: Autonomous 3-D positioning of surgical instruments in robotized laparoscopic surgery using visual servoing. IEEE Trans. on Robo. and Auto. 19(5), 842–853 (2003)
9. Tonet, O., Armes, T.U., Magali, G., Dario, P.: Tracking Endoscopic Instruments Without Localizer: Image Analysis-Based Approach Stud. Health Tech. Inf. 119, 544–549 (2005)
10. Cano, A.M., Lamata, P., Gayá, F., del Pozo, F., Gómez, E.J.: New Methods for Video-Based Tracking of Laparoscopic Tools. In: Harders, M., Székely, G. (eds.) ISBMS 2006. LNCS, vol. 4072, pp. 142–149. Springer, Heidelberg (2006)
11. Voros, S., Orvain, E., Cinquin, P.: Automatic detection of instruments in laparoscopic images: a first step towards high level command of robotized endoscopic holders. In: BIOROB 2006 1st IEEE/RAS-EMBS Int. Conf. On Biomed. Rob. and Biomecha., Pisa (2006)

Coronary Motion Modelling for Augmented Reality Guidance of Endoscopic Coronary Artery Bypass

Michael Figl[1], Daniel Rueckert[1], David Hawkes[3], Roberto Casula[4],
Mingxing Hu[3], Ose Pedro[1], Dong Ping Zhang[1], Graeme Penney[5],
Fernando Bello[2], and Philip Edwards[1,2]

[1] Department of Computing, Imperial College London, UK
[2] Department of Biosurgery and Surgical Technology, Imperial College London, UK
[3] Centre of Medical Image Computing, University College London, UK
[4] Cardiothoracic Surgery, St. Mary's Hospital, London, UK
[5] Division of Imaging Sciences, King's College London, UK

Abstract. The overall aim of our project is to guide totally endoscopic coronary artery bypass. This requires construction of a 4D preoperative model of the coronary arteries and myocardium. The model must be aligned with the endoscopic view of the patient's beating heart and presented to the surgeon using augmented reality. We propose that the model can be constructed from coronary CT. Segmentation can be performed for one phase of the cardiac cycle only and propagated to the others using non-rigid registration. We have compared the location of the coronaries produced by this method to hand segmentation.

Registration of the model to the endoscopic view of the patient is achieved in two phases. Temporal registration is performed by identification of corresponding motion between model and video. Then we calculate photo-consistency between the two da Vinci endoscope views and average over the frames of the motion model. This has been shown to improve the shape of the cost function. Phantom results are presented.

The model can then be transformed to the calibrated endoscope view and overlaid using two video mixers.

1 Introduction

1.1 Augmented Reality Applications in Surgery

With the introduction of the da Vinci robot, totally endoscopic coronary artery bypass grafting (TECAB) can be performed on the beating heart. This provides less invasive surgery without the need for sternotomy or heart-lung bypass. However, the level of conversion to more invasive endoscopic procedures is reported to be 20-30% with the most likely causes being misidentification of the target coronary artery or difficulty in locating the vessel [1,2].

To investigate whether augmented reality (AR) guidance can improve TECAB, we have identified two critical points in the procedure. During harvesting of the

F. Bello, E. Edwards (Eds.): ISBMS 2008, LNCS 5104, pp. 197–202, 2008.

left internal mammary artery highlighting the position of the bifurcation would allow surgery to progress rapidly to this point. After opening of the pericardium overlay of the target vessel would allow accurate location and identification. Accurate coronary vessel overlay requires alignment of a 4D preoperative model of the beating heart to the da Vinci endoscope view and AR visualisation.

2   Methods

To achieve guidance we require a 4D model of the beating heart. This must be both temporally and spatially registered to the patient. Finally the model must be visualised using the calibrated endoscope view.

2.1   4D Model Construction

The model of the patient's beating heart comes from coronary CT, which can be reconstructed at up to 20 multiple even phases throughout the cardiac cycle. The relevant vessels must be segmented along with the surface of the myocardium.

The motion of the heart can potentially be obtained from non-rigid image registration [3]. This means that the heart and coronaries need only be extracted in one frame and the shape in subsequent frames can be propagated by registration [4]. This has been demonstrated to give good results for the surface of the heart in animal studies [3]. However, we have found that the position of the coronaries is not always well predicted by registration. This could be due to the fact that though the surface is well aligned, cardiac twisting motion or sliding along the surface may not be detected by registration (see Figure 1).

A subdivision surface representation of the heart with volume preservation has been investigated to track the motion of the myocardium [5]. We have considered whether vessel enhancement, automated presegmentation of the CT data-set or the use of a 4D statistical model could be used to improve segmentation [6]. An example of a patient model can be seen in figure 4(a).

2.2   Registration Techniques

Having obtained a preoperative model of the patient we now need to align this model to the video view through the da Vinci endoscope.

Our strategy in performing registration will be to separate the temporal and spatial alignment of the preoperative model. Temporal alignment may be obtained using the ECG signal. However, there may be residual lag between the ECG and the video and we are investigating whether feature tracking in the video images could be used for this purpose.

Corresponding features in the left and right views in the da Vinci stereo-endoscope can provide 3D tracked points on the viewed surface. We propose this technique to measure the motion of the heart and to separate cardiac and respiratory motion. We are also examining whether geometry constraints can be used as an accurate means of finding the period of the heart cycle, in the presence of rigid motion of the camera or near rigid motion due to breathing.

Having established temporal registration, the remaining motion should be rigid. Preliminary work has shown that 3D-3D non-rigid registration can be used to build separate models of respiratory motion of the liver [7] and heart [8]. Since both respiratory and cardiac motion are smooth and continuous, we will develop 4D parametric motion models based on 4D B-splines.

To establish correspondence we are investigating two approaches. We are adopting a similar approach to that of [9] to reconstruct the motion of the viewed surface, which can then be registered to the preoperative 4D model. Secondly we are investigating whether intensity-based similarity metrics like photo-consistency [10] can be used.

2.3   Visualisation

In order to provide an AR facility for the stereo endoscope of the da Vinci system we need to establish the relationship between camera coordinates and image coordinates (determined by the intrinsic transformation parameters) as well as the relationship between world coordinates and camera coordinates (determined by the extrinsic transformation parameters). For camera calibration we use an offline camera calibration technique to determine these parameters [11].

As the image fusion is done by use of chroma keying with two video mixers, an additional 2D-2D transformation from the graphical output to the input channels of the mixer is needed. This is achieved by point to point registration.

We will then use the model-based 2D-3D registration algorithm described previously to estimate the extrinsic transformation parameters.

3   Results

3.1   4D Model Construction

For preoperative model building we use coronary CT images reconstructed at 20 even phases throughout the cardiac cycle. We segment the first phase by hand and propagate this to the other 19 phases by non-rigid registration [4]. This is similar to [3], but here it is demonstrated on clinical CT scans rather than animal data. Figure 1 gives an idea of the quality of the registration.

3.2   Registration to the Endoscopic View

Figure 2 shows the behaviour of the photo-consistency measure near the minimum for video images from a bench camera setup. As the number of phases of the phantom CT surface model used increases, the curve becomes smoother. This suggests that our strategy of performing rigid spatial registration but averaging temporally corresponding frames is sound. The resulting alignment using photo-consistency can be seen in figure 3. An initial registration is performed using manual point identification on the surface model and in both video images. The model is aligned iteratively so as to minimise the 2D projection error of the landmarks, which was found to be more robust than reconstructing 3D points

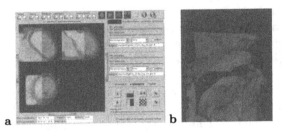

a   b

Fig. 1. Non-rigid registration was used to register the first image to the other images. **a** shows the registered image slices from phases 0% and 60 % of the cardiac cycle; **b** shows an automated segmentation of the myocardium with aligned manual and registration-based vessel segmentations. Some discrepancy between the true vessel position and that from registration is seen.

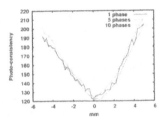

Fig. 2. Graph showing photo-consistency vs displacement near the minimum using 1, 5 and 10 phases of the phantom surface model. Noise and local minima are averaged out as more phases are used.

from the two video images and performing a 3D registration. There are still some inaccuracies as can be seen in figure 3(a). After photo-consistency registration the alignment is seen to be much better (figure 3(b)).

3.3 Visualisation

An example of retrospectively aligned images is shown in figure 4. We have a number of clinical coronary CT images that are being used to investigate preoperative model construction. Registration algorithms using both feature tracking and intensity-based techniques are being developed.

4 Discussion

We propose a system for augmented reality image guidance of totally endoscopic coronary artery bypass. Previous work has suggested the use of image guidance in TECAB surgery and demonstrated its feasibility on a heart phantom [12]. Falk et al have demonstrated a system for AR guidance based on multi-planar x-ray angiography [13]. We describe the first such results using clinical coronary CT scans to provide the 4D patient model using non-rigid registration. We

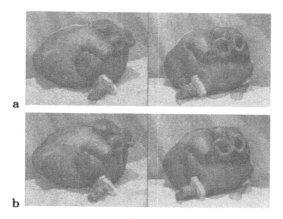

Fig. 3. Photo-consistency registration result from a bench camera setup, showing **a** the initial position from manual point-based registration and **b** the result of photo-consistency alignment. A clear improvement is achieved using photo-consistency.

Fig. 4. A rendering of the preoperative model showing the myocardial surface, left internal mammary artery, left anterior descending artery and a diagonal branch (a), an aligned rendering (b) and its corresponding endoscope view (c)

also propose two novel strategies for alignment of this model to the endoscopic view. The first uses robust feature tracking to reconstruct the viewed surface, which can then be matched to the preoperative model. The second strategy uses intensity-based methods for registration. The latter has been demonstrated on a phantom images.

For augmentation of the endoscopic view we use chroma-keying by video mixers , which does not introduce any lag to the surgeons view of the real surface. It is hoped that such information can improve the efficiency of TECAB surgery and reduce the conversion rate to more invasive procedures.

Acknowledgments

We would like to thank the EPSRC for funding this project. We are also grateful to the theatre staff at St Mary's hospital, London and to the radiology staff in St Mary's and the Royal Brompton hospitals for their cooperation.

202 M. Figl et al.

References

1. Dogan, S., Aybek, T., Andressen, E., Byhahn, C., Mierdl, S., Westphal, K., Matheis, G., Moritz, A., Wimmer-Greinecker, G.: Totally endoscopic coronary artery bypass grafting on cardiopulmonary bypass with robotically enhanced telemanipulation: Report of forty-five cases. J. Thorac. Cardiovasc. Surg. 123, 1125–1131 (2002)
2. Falk, V., Diegeler, A., Walther, T., Banusch, J., Brucerius, J., Raumans, J., Autschbach, R., Mohr, F.W.: Total endoscopic computer enhanced coronary artery bypass grafting. Eur. J. Cardio-Thorac. Surg. 17, 38–45 (2000)
3. Wierzbicki, M., Drangova, G., Guiraudon, G., Peters, T.M.: Validation of dynamic heart models obtained using non-linear registration for virtual reality training, planning, and guidance of minimally invasive cardiac surgeries. Med. Image Anal. 8, 387–401 (2004)
4. Rueckert, D., Sonoda, L.I., Hayes, C., Hill, D.L.G., Leach, M.O., Hawkes, D.J.: Nonrigid registration using free-form deformations: Application to breast MR images. IEEE Trans. Med. Imaging 18(8), 712–721 (1999)
5. Chandrashekara, R., Mohiaddin, R., Razavi, R., Rueckert, R.: Nonrigid image registration with subdivision lattices: Application to cardiac mr image analysis. In: Taylor, C., Colchester, A. (eds.) MICCAI 1999. LNCS, vol. 1679, pp. 335–342. Springer, Heidelberg (1999)
6. Perperidis, D., Mohiaddin, R., Edwards, P., Rueckert, D.: Hill: Segmentation of cardiac MR and CT image sequences using model-based registration of a 4D statistical model. In: Proc. SPIE Medical Imaging 2007, vol. 6512 (2007)
7. Blackall, J.M., Penney, G.P., King, A.P., Hawkes, D.J.: Alignment of sparse freehand 3-d ultrasound with preoperative images of the liver using models of respiratory motion and deformation. IEEE Trans. Med. Imaging 24, 1405–1416 (2005)
8. McLeish, K., Hill, D.L.G., Atkinson, D., Blackall, J.M., Razavi, R.: A study of the motion and deformation of the heart due to respiration. IEEE Trans. Med. 21, 1142–1150 (2002)
9. Stoyanov, D., Mylonas, G.P., Deligianni, F., Darzi, A., Yang, G.Z.: Soft-tissue motion tracking and structure estimation for robotic assisted mis procedures. In: Duncan, J.S., Gerig, G. (eds.) MICCAI 2005. LNCS, vol. 3750, pp. 139–146. Springer, Heidelberg (2005)
10. Clarkson, M.J., Rueckert, D., Hill, D.L.G., Hawkes, D.J.: Using photo-consistency to register 2d optical images of the human face to a 3D surface model. IEEE Trans. Pattern Anal. Mach. Intell. 23, 1266–1280 (2001)
11. Bouguet, J.: Camera calibration toolbox for matlab (2007), http://www.vision.caltech.edu/bouguetj
12. Szpala, S., Wierzbicki, M., Guiraudon, G., Peters, T.M.: Real-time fusion of endoscopic views with dynamic 3-d cardiac images: A phantom study. IEEE Trans. Med. Imaging 24, 1207–1215 (2005)
13. Falk, V., Mourgues, F., Adhami, L., Jacobs, S., Thiele, H., Nitzsche, S., Mohr, F.W., Coste-Maniere, T.: Cardio navigation: Planning, simulation, and augmented reality in robotic assisted endoscopic bypass grafting. Ann. Thorac. Surg. 79, 2040–2048 (2005)

Quantifying Effects of Class I Anti-arrhythmic Drugs in Human Virtual Cardiac Tissues

Alan P. Benson[1], Jennifer A. Lawrenson[2], Stephen H. Gilbert[1], and Arun V. Holden[1]

[1] Institute of Membrane and Systems Biology & Multidisciplinary Cardiovascular Research Centre, University of Leeds, Leeds LS2 9JT, UK
a.p.benson@leeds.ac.uk
http://www.cbiol.leeds.ac.uk
[2] School of Medicine, University of Leeds, Leeds LS2 9JT, UK

Abstract. Ventricular tachycardia and fibrillation are dangerous cardiac arrhythmias. Anti-arrhythmic drugs such as lidocaine (a class I drug) reinstate normal sinus rhythm, but they have pro-arrhythmic side effects and their mechanisms of action are poorly understood. We used computational models of human ventricular tissues to elucidate and quantify these mechanisms.

Keywords: Cardiac tissue, Computational modelling, Class I anti-arrhythmic drugs, Re-entry.

1 Introduction

Ventricular tachycardia (VT) and fibrillation (VF) are dangerous cardiac arrhythmias. Anti-arrhythmic drugs such as lidocaine – a class Ib drug that blocks the sodium channel – can prevent such arrhythmias. They exert their anti-arrhythmic effects by reducing the conduction velocity of propagating excitation and by increasing effective refractory period, which reduces cell excitability and prevents excessive repetitive depolarisation. Whilst these are desirable effects, they also increase the potential of the drug to become pro-arrhythmic. However, the mechanisms underlying these side effects are poorly understood [1]. Virtual tissues provide a means for visualising, dissecting and explaining propagation in the normal and pharmacologically-modified heart. We used a hierarchy of human virtual ventricular cells and tissues to identify and quantify both the anti- and pro-arrhythmic effects of class I anti-arrhythmics.

2 Methods

For a single cell, the rate of change of membrane potential V is given by

$$dV/dt = -I_{\mathrm{ion}}, \tag{1}$$

F. Bello, E. Edwards (Eds.): ISBMS 2008, LNCS 5104, pp. 203–208, 2008.

where I_{ion} is the membrane current density. For I_{ion} we used the model of Ten Tusscher & Panfilov [2], which provides formulations for endocardial, midmyocardial and epicardial cells. The effects of lidocaine were simulated using the guarded receptor model, with the drug binding to the inactivated state of the sodium channel [3]. The sodium current (I_{Na}) formulation from [2] then becomes

$$I_{Na} = G_{Na}m^3hj(1-b)(V-E_{Na}) \ , \qquad (2)$$

$$db/dt = K(1-h)(1-b) - Ib \ , \qquad (3)$$

where b is the fraction of blocked channels, and $K = 31.0 \,/mM/s$ and $I = 0.79 \,/s$ are binding and unbinding rates respectively, as measured experimentally in human ventricular cells [4]. This formulation causes the drug to bind and unbind to the sodium channel during the course of a single action potential (see Fig. 1), and allows the rate-dependency of the drug to be modelled (see Results).

The single cell drug effects were incorporated into a hierarchy of virtual ventricular tissues: (i) a 15 mm 1D heterogeneous transmural strand; (ii) a 12×12 cm 2D epicardial sheet with anisotropic propagation (i.e. fibre orientations); and (iii) a 3D heterogeneous wedge model of the left ventricular wall with orthotropic architecture (fibre and sheet orientations) obtained from diffusion tensor magnetic resonance imaging (DT-MRI) – see [5] and Acknowledgments for details. Propagation of electrical excitation in these virtual tissues was described by the non-linear cable equation

$$\partial V/\partial t = \nabla(\mathbf{D}\nabla V) - I_{ion} \ , \qquad (4)$$

where ∇ is a spatial gradient operator and $\mathbf{D}$ is the diffusion tensor that characterises electrotonic spread of voltage through the tissue. Parameter values and integration details for equations (1) and (4) can be found in reference [5].

3   Results

3.1   Model Characterisation

Figure 2 and Table 1 show that simulated application of 0.08 mM lidocaine produced rate-dependent steady state (end systolic) block of I_{Na} and reductions in maximum action potential upstroke velocity dV/dt_{max}, with the percentage of block and corresponding degree of dV/dt_{max} reduction being greatest with faster pacing. There was little effect on action potential duration (APD). These results are consistent with experimental findings [6]. Conduction velocity restitution in a 1D strand model of untreated tissue and with simulated application of 0.08 mM lidocaine is shown in Fig. 3. Tissue was paced from the endocardial end of the strand until a steady state was achieved. The reduced dV/dt_{max} under drug conditions (see the single cell results in Fig. 2b) caused reduced conduction velocity compared to untreated tissue, and propagation block at cycle lengths shorter than 440 ms.

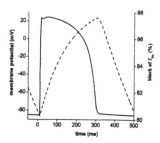

Fig. 1. The action potential $V(t)$ (solid line) and percentage of I_{Na} block due to drug binding (dashed line) in an epicardial cell model paced at 2 Hz

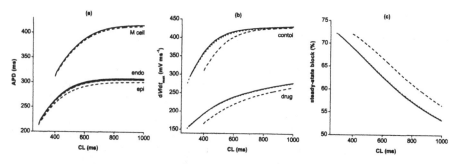

Fig. 2. Restitution in virtual cells, either untreated or with the simulated application of 0.08 mM lidocaine. (a) APD in endocardial (endo), midmyocardial (M cell) and epicardial (epi) control cells (dashed lines) and with 0.08 mM lidocaine (solid lines). (b) dV/dt_{max} in endocardial (solid), midmyocardial (dashed) and epicardial (dotted) control cells and with 0.08 mM lidocaine. (c) Steady-state end systolic block of I_{Na} in endocardial (solid), midmyocardial (dashed) and epicardial (dotted) cells with 0.08 mM lidocaine. CL, cycle length.

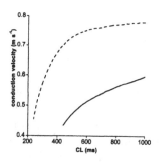

Fig. 3. Conduction velocity restitution in a 1D strand model of untreated tissue (dashed line) and with 0.08 mM lidocaine (solid line)

Table 1. Effects of simulated application of 0.08 mM lidocaine in single virtual cells at 1 and 2 Hz pacing. All values are percentages.

	1 Hz			2 Hz		
	I_{Na} block	$\Delta dV/dt_{max}$	ΔAPD	I_{Na} block	$\Delta dV/dt_{max}$	ΔAPD
endocardial	53.2	−35.6	+1.7	65.9	−44.9	+2.5
midmyocardial	56.4	−38.0	+0.6	69.6	−46.9	+0.7
epicardial	53.3	−35.6	+0.8	66.0	−44.7	+0.9

3.2 Anti-arrhythmic Mechanism

The simulations in this section all used the "standard" parameter set from [2] that causes APD alternans at short cycle lengths, and therefore breakdown of a rapidly rotating re-entrant wave (i.e. VT) into the chaotic behaviour seen during VF. In 2D virtual tissues, the reduced conduction velocity (see Fig. 3) prevented initiation of VT, prevented breakdown of VT into VF, or shortened the lifetime of the arrhythmia, depending on simulated drug concentration. The 2D virtual tissues were paced 10 times at 2.5 Hz (normal sinus rhythm, NSR) before VT was initiated. In untreated tissue, VT quickly broke down into VF, and persisted for at least 10 s (see Fig. 4a). As the tissue progressed from NSR through to VT and then VF, the rate of excitation increased, and so cycle length became shorter. With the simulated application of 0.015 mM lidocaine, VT degenerated into VF as in the untreated tissue, but the VF died out after approximately 2.5 s (see Fig. 4b) due to the short cycle length seen during VF resulting in increased binding of the drug (see Fig. 2c), and therefore reduced excitability and conduction velocity. With a higher concentration of 0.08 mM lidocaine, VT was stable and no breakdown into VF occurred (see Fig. 4c). In this situation, conduction velocity during VT was reduced to such an extent that the cycle length remained on the stable part of the APD restitution curve (see [2]), and thus APD alternans never appear. For lidocaine concentrations greater than 0.5 mM, 2:1 conduction block occurred during NSR. Qualitatively similar results are seen in the 3D orthotropic and heterogeneous left ventricular free wall wedge model (see Fig. 4d-e).

3.3 Pro-arrhythmic Mechanism

The vulnerable window identifies the spatio-temporal location where an ectopic excitation produces unidirectional propagation. In 2D and 3D, such unidirectional propagation can initiate a pair of re-entrant waves, and the 1D strand model can be used to quantify the vulnerable window. In our simulations, the reduced conduction velocity quantified in the 1D strand model (see Fig. 3) was associated with an increased vulnerable window, caused by increased dispersion of repolarisation (see Fig. 5). However, this increased dispersion of repolarisation is not due to heterogeneous increases in APD (seen, for example, with class III drugs [7]) but due to reduced conduction velocity causing epicardial (but not endocardial) tissue to depolarise, and therefore repolarise, much later than under control conditions. The maximum temporal width of the vulnerable window,

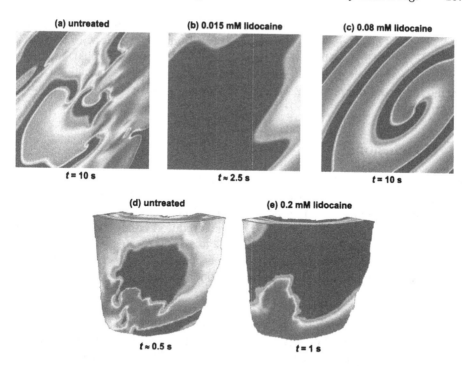

Fig. 4. Top: Snapshots showing effects of lidocaine on VT and VF in 2D virtual tissues. Excited tissue is red/lighter, resting tissue is blue/darker. See text for details. Bottom: Snapshots showing effects of lidocaine on VT and VF in the 3D left ventricular free wall wedge model. The snapshot of the untreated tissue shows the beginning of re-entrant spiral wave breakup (i.e. the transition from VT to VF). In tissue with 0.2 mM lidocaine, VT remains stable and the breakdown into VF does not occur.

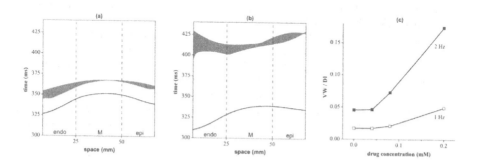

Fig. 5. Vulnerable windows in 1D strand models are shown for (a) untreated tissue paced at 1 Hz and (b) tissue with 0.2 mM lidocaine paced at 2 Hz. Both panels show time of repolarisation through the strand (solid black line) and the spatiotemporal extent of the vulnerable window (grey area). (c) The maximum temporal width of the vulnerable window, expressed as a fraction of the diastolic interval, increased with pacing rate and simulated drug concentration.

expressed as a percentage of the diastolic interval, increased with pacing rate and simulated drug concentration (see Fig. 5c), from 1% at 1 Hz with 0.015 mM lidocaine, to 18% at 2 Hz with 0.2 mM lidocaine. We conclude that simulated application of class I drugs increases the vulnerable window, especially at faster pacing rates, and therefore increases the probability of any ectopic beat initiating an arrythmia.

4 Conclusions

Using our models of human virtual ventricular cells and tissues, along with pharmacological modifications where lidocaine binds to the inactivated state of the sodium channel in a rate-dependent manner, we have shown that the anti-arrhythmic mechanism of class I drugs is reduced conduction velocity and excitability, and therefore propagation block at the high rates of excitation seen during arrhythmias. The pro-arrhythmic mechanism is an increase in the vulnerable window for unidirectional propagation, also caused by the reduced conduction velocity.

Acknowledgments. We thank Drs. P.A. Helm and R.L. Winslow at the Center for Cardiovascular Bioinformatics and Modeling and Dr E. McVeigh at the National Institute of Health for the human DT-MRI dataset. This work was supported by the European Union through the BioSim Network of Excellence (contract No. LSHB-CT-2004-005137), The Dr Hadwen Trust for Humane Research, Heart Research UK and the Medical Research Council.

References

1. Chaudhry, G.M., Haffajee, C.I.: Anti-arrhythmic agents and pro-arrhythmia. Crit. Care Med. 28, N158–N164 (2000)
2. Ten Tusscher, K.H.W.J., Panfilov, A.V.: Alternans and spiral breakup in a human ventricular tissue model. Am. J. Physiol. 291, H1088–H1100 (2006)
3. Starmer, C.F., Lastra, A.A., Nesterenko, V.V., Grant, A.O.: Pro-arrhythmic response to sodium channel blockade. Circulation 84, 1364–1377 (1991)
4. Furukawa, T., Koumi, S.-I., Sakakibara, Y., Singer, D.H., Jia, H., Arentzen, C.E., Backer, C.L., Wasserstrom, J.A.: An analysis of lidocaine block of sodium current in isolated human atrial and ventricular myocytes. J. Mol. Cell. Cardiol. 27, 831–846 (1995)
5. Benson, A.P., Halley, G., Li, P., Tong, W.C., Holden, A.V.: Virtual cell and tissue dynamics of ectopic activation of the ventricles. Chaos 17, 015105 (2007)
6. Weirich, J., Antoni, H.: Rate-dependence of anti-arrhythmic and pro-arrhythmic properties of class I and class III anti-arrhythmic drugs. Basic Res. Cardiol. 93, 125–132 (1998)
7. Benson, A.P., Aslanidi, O.V., Zhang, H., Holden, A.V.: The canine virtual ventricles: a platform for dissecting pharmacological effects on propagation and arrhythmogenesis. Prog. Biophys. Mol. Biol. 96, 187–208 (2008)

Development of a Microscope Embedded Training System for Neurosurgery

Alessandro De Mauro[1], Jörg Raczkowsky[1], Reiner Wirtz[2], and Heinz Wörn[1]

[1] Institute for Process Control and Robotics, University of Karlsruhe (TH) , Germany
[2] Neurosurgical Department, Clinic University of Heidelberg, Germany
demauro@ira.uka.de

Abstract. Virtual reality based training systems for surgery, have recently shown great potential as an alternative to traditional training methods. In neurosurgery, state of art training devices is limited to a few examples. They are based either on traditional displays or head-mounted displays. The aim of this research is the development of the first virtual reality training system for neurosurgical interventions based on a real surgical microscope for a better visual and ergonomic realism. The simulation takes advantage of an accurate tissue modeling, a force feedback device and a rendering of the virtual scene directly to the oculars of the operating microscope. A prototype of a stereoscopic Augmented Reality microscope for intra-operative presentation of preoperative three-dimensional data has been realized in our laboratory. We are reusing the image injection component of this existing platform developing a training system for educational and preoperative purposes based on virtual organs reconstructed from real patients' images.

Keywords: Virtual Reality, Training Systems, Visualization, Image guided therapy, Neurosurgery.

1 Introduction

Currently, the outcome of any intervention is closely related to the surgeon's skills. Neurosurgery requires detailed knowledge of the anatomy of the central nervous system, and demands great manual dexterity for surgical work within the brain and spinal cord. In fact, wrong operative movements can be devastating leaving patients paralyzed, comatose or dead. Traditional techniques for training in surgery use animals, phantoms, cadavers and real patients. Especially in neurosurgery, minimally invasive techniques require different types training and today there is a new possibility: virtual reality. It has shown that surgical residents trained to perform surgery using virtual reality simulators are more proficient and have fewer errors in the first operations than those who received no virtual reality simulated education [1]. In other words, virtual reality simulators help to accelerate the learning curve. Surgical simulators offer sophisticated task training programs, record errors and provide a way of measuring operative efficiency and performance, working both as an educational tool and a skills

F. Bello, E. Edwards (Eds.): ISBMS 2008, LNCS 5104, pp. 209–214, 2008.

validation instrument. This scientific work is focused on developing a neurosurgical simulator directed towards both educational and preoperative purposes based on a virtual environment set up on reconstructed human organs from real patients' images and using a real operating microscope (Carl Zeiss).

2 Background in Surgical Simulation

The current state-of-the-art technology in the field of surgical training system is based prevalently on: the LapSim system (SurgicalScience) [2], Endoscopy AccuTouch System (Immersion) [3] and KISMET (Forschungszentrum, Karlsruhe) [4]. These examples are related to different endoscopic techniques (laparoscopy, bronchoscopy, gynecology, etc...). They simulate common procedures in minimal invasive surgery carried out during real surgery, through natural body openings or small incisions. On the other hand, in the state-of-the-art in neurosurgical simulation [5], shows only a few examples of VR based systems which use force feedback [6, 7, 8]. Because a neurosurgical microscope is used in a large percentage of interventions, the realism of these simulators is limited by the architecture: they use either displays or head mounted displays and not a real surgical microscope; in addition, an operating microscope is a very expensive resource and is not used for normal standard training but only for the real operations.

In a previous work a prototype stereoscopic Augmented Reality (AR) microscope for ergonomic intra-operative presentation of complex preoperative three-dimensional data for neurosurgical interventions was realized at our institute [9].

We are reusing a part of this existing AR-platform (the image injection component) converting it to a VR system.

The development of the very first virtual reality training system for neurosurgical interventions based on a real operating microscope and haptic feedback is the aim of this scientific work.

3 Method and Tools

In the operating theater, in order to understand the correct tumor position compared to the preoperative images (CT, MRI) surgeon's eyes are normally on the microscope oculars and occasionally surgeons look at the screen. At these considerations and in order to have a complete training system, it is requested to simulate the virtual view inside the microscope, to provide the user's hand with force feedback and to simulate the navigational software actually used in OR (i.e. BrainLab or Stryker). Fig.1 and 2 show simulator concepts and the platform. The building of a virtual environment using real patients' data from medical image devices and interaction with such an environment helps students in their understanding of the basic human anatomy. For this reason human organs are accurately reconstructed from real patient images using the tool 3DSlicer [10]. This software has been used routinely for segmentation, classification, and as an intraoperative navigational tool for biopsies and open craniotomies. The 3D objects obtained by 3DSlicer segmentation are imported directly to our application. In order to provide the 3D models of the surgical tools we are using a Laser ScanArm (FARO) [11]. Fig. 3 describes the simulation software architecture. The rendering software developed in C++ and Python is built on the open source GPU licensed and cross-platform H3D [12], a scene-graph API based on OpenGL for graphics rendering, OpenHaptics for haptic rendering and X3D [13] for the 3D environment description (Fig. 4).

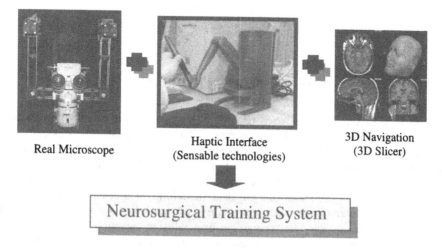

Real Microscope

Haptic Interface
(Sensable technologies)

3D Navigation
(3D Slicer)

Neurosurgical Training System

Fig. 1. Simulator concepts

Fig. 2. Simulator architecture

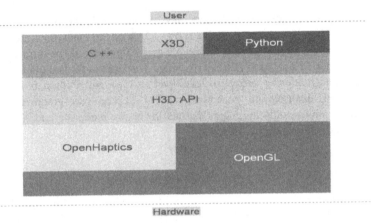

Fig. 3. Simulation software architecture

A tracking system infrared based (Polaris NDI) provides the real-time position of the two viewpoints (one of each microscope ocular) in the virtual environment using passive tools positioned on the microscope. The collisions between organs and surgical tools produce forces which have to be replicated by the haptic interface and organ deformations, which have to be graphically rendered (Fig. 5). We are developing collision detection and physical modelling modules using OpenGL in order to obtain better performance and realism of the deformations. Hierarchies of bounding volumes provide a fast way to perform collision detection between complex models and the method used is based on model partition [14], the strategy of subdividing a set of objects into geometrically coherent subsets and computing a bounding volume for each subset of objects. It's a fast method for collision query, which can be successfully used for object's deformation. The force computation and the organ deformation are dependent on a physical model describing the mechanical properties of the virtual bodies. Deformable object modeling has been studied, in computer graphics and animation, for many years. Some results shows that brain retraction and resection can be well modeled [15]. Accurate physically based methods are used in the medical simulation field to represent human tissues: Finite Element Method (FEM), Long Element Method (LEM), and Mass Spring Damper (MSD). FEM and LEM are continuous models, mathematically robust with a high level of realism and a high computational load. MSD method consists of a mesh of point masses connected by elastic links and mapped onto the geometric representation of the virtual object. In addition, it's a discrete method characterized by low computable load, simplicity, low accuracy and risk of instability because it uses Newton dynamic to modify the point-masses positions and creates deformations considering the volume conservation. We are using MSD to model the brain tissue evaluating and comparing the use of FEM. In order to have a complete training platform we need a video navigation system containing different 3D volume views. To achieve this, we have connected our system with the Image Guided Therapy module of 3DSlicer. We are using a haptic device (Phantom Desktop) in order to provide the surgeon with an immersive experience during the interaction between the surgical tools and the brain or skull of the virtual patients. The force feedback workspace and other important properties (ex. stiffness range and nominal position resolution) make it suitable to be used in an ergonomic way together with the microscope. All the basic surgical operations on the brain are suitable to increase surgery skills in a fast and safe way using virtual reality. The first operating task we decided to simulate are the visual and tactile sensations of brain palpation (pushing aside the human tissue using a neurosurgical spatula) and a correct patient positioning because it is such a key factor for the operation. We are also evaluating together with our medical partner (Neurosurgical Department of Clinic University of Heidelberg), the possibility in simulating other more difficult tasks (i.e. craniotomy). Other future improvements are concerned with the augmented reality concepts: adding patient registration, microscope (oculars) calibration and extending the optical tracking to patient (phantom) and surgical tools with these concepts we will propose a new augmented reality microscope for intraoperative use. The choice of X3D for the environment description will permit a future natural extension of the work as a distributed training platform using web technologies.

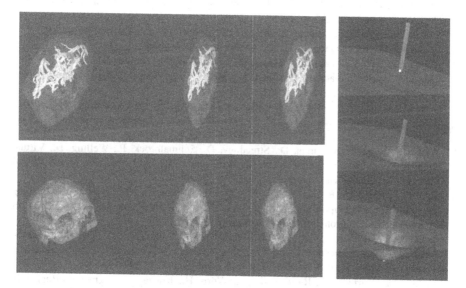

Fig. 4. Different simulator scenes rendered on screen (left images) and in microscope oculars (right images). 3D different organs patient reconstructed from real CT images (ventricles, vessels, skull and brain). The oculars view is stereoscopic.

Fig. 5. Deformations of the Soft tissue during the interaction with a surgical tool

4 Conclusions

We are developing the real first neurosurgical simulator based on a real microscope. The architecture and the main features were defined in tight collaboration with surgeon staff. Force feedback interaction with soft and hard tissues is provided to the surgeon's hands in order to complete the immersive experience. A complex and accurate virtual environment is provided, by a stereoscopic image injection directly inside the operating microscope. The navigation inside the real patient's images is shown on the screen, using 3DSlicer directly connected to the tracking system. The use of the previously described software architecture is a guarantee of performance and portability. All the components are open source or at least based on a GPL license. Future work will be focused on the simulation of craniotomy and the tumor resection and more difficult tasks. Currently we are working to improve realistic deformations and tactile sensations providing ad-hoc collision detection and physical modeling. We are also evaluating to extend the work to a distributed training.

Acknowledgments. This research is a part of the European project "CompuSurge" funded by "Marie Curie" research network.

References

1. Ro, C.Y., et al.: The LapSim: A Learning Environment for Both Experts and Novices, Department of Surgery, St. Luke's-Roosevelt Hospital Center and Columbia University, New York, New York, U.S.A (2005)
2. SurgicalScience (Status January 2008), http://www.surgical-science.com/
3. Immersion (Status January 2008), http://www.immersion.com/
4. Kühnapfel, U., Çakmak, K., Maass, H., Waldhausen, S.: Models for simulating instrument-tissue interactions, 9th Medicine Meets Virtual Reality 2001, Newport Beach, CA, USA (2001)
5. Goh, K.Y.C.: Virtual Reality Application in Neurosurgery. In: Proceedings of the 2005 IEEE, Engineering in Medicine and Biology 27th Annual Conference, Shanghai, China (2005)
6. Wiet, G., Bryan, J., Sessanna, D., Streadney, D., Schmalbrock, P., Welling, B.: Virtual temporal bone dissection simulation. In: Westwood, J.D. (ed.) Medicine Meets Virtual Reality, Amsterdam, The Netherlands, pp. 378–384 (2000)
7. Sato, D., Kobayashi, R., Kobayashi, A., Fujino, S., Uchiyama, M.: Soft Tissue Pushing Operation Using a Haptic Interface for Simulation of Brain Tumor Resection. Journal of Robotics and Mechatronics 18, 634–642 (2006)
8. Luciano, C., Banerjee, P., Lemole, M.G., Charbel, F.:Second Generation Haptic Ventriculostomy Simulator Using the ImmersiveTouch<Superscript>TM</Superscript> System, Medicine Meets Virtual Reality 14, Long Beach, CA (2008)
9. Aschke, M., Wirtz, C.R., Raczkowsky, J., Wörn, H., Kunze, S.: Augmented Reality in Operating Microscopes for Neurosurgical Interventions. In: Wolf, L.J., Strock, J.L. (eds.) 1st International IEEE EMBS Conference on Neural Engineering, pp. 652–655 (2003)
10. 3DSlicer (status January 2008), http://www.slicer.org/
11. Faro (status January 2008), http://www.faro.com
12. H3D (status January 2008), http://www.h3dapi.org/
13. X3D (status January 2008), http://www.web3d.org/x3d/
14. Van den Bergen, G.: Collision Detection in Interactive 3D Environment. Elsevier Morgan Kaufmann Publishers, San Francisco (2004)
15. Miga, M.I., Roberts, D.W., Kennedy, F.E., Platenik, L.A., Hartov, A., Lunn, K.E., Paulsen, K.D.: Modeling of retraction and resection for intraoperative updating of images during surgery. In: Neurosurgery, vol. 49(1). Lippincott Williams & Wilkins, Hagerstown (2001)

A Virtual Environment for Core Skills Training in Vascular Interventional Radiology

Vincent Luboz, James Lai, Rafal Blazewski, Derek Gould, and Fernando Bello

Biosurgery and Surgical Technology Department, SORA, Imperial College London – St Mary's Hospital, London, QEQM, W2 1NY, UK
{vluboz,fernando}@imperial.ac.uk
http://www1.imperial.ac.uk/medicine/people/v.luboz/

Abstract. We present a prototype for a new virtual environment aiming at training interventional radiologists in the core skills involved in using catheters or guidewires. The instrument is modelled as a hybrid mass-spring particle system while the vasculature is a rigid triangulated surface mesh. A specially designed commercial haptic device allows the trainee to use real instruments to guide the simulation through synthetically generated vasculature with different degrees of complexity. Open source libraries are used for the visualisation and collision detection. Preliminary results show reasonable behaviour.

Keywords: hybrid mass-spring model, haptic, interventional radiology.

1 Introduction

This paper proposes an alternative for core skills training in vascular Interventional Radiology (IR). IR uses medical imaging to guide minimally invasive techniques involving needles, wires and catheters to access and treat vascular structures. These manipulations require a high level of skill to ensure a safe and positive outcome. The current training method relies on traditional apprenticeships, which are widely acknowledged to be expensive, risky and increasingly inadequate for training

Several alternative training methods making use of computer-based simulation have been proposed to improve the training of the clinicians. Such virtual environments offer a safe framework to learn and practice the specific skills of IR without putting patients at risk. Furthermore, trainees are free to practice at their own pace and are given useful and objective feedback on their performance, thus quantifying their procedural competency. Several companies, such as Mentice and Simbionix, have commercial vascular IR simulators offering a range of vascular cases and pathologies. However, none of the existing systems focus on core or basic skill training, which has been shown to be of great importance in laparoscopic surgery [1].

Catheter and guidewire modelling is a key component of a simulator for vascular IR, with several research groups exploring different approaches aiming at improving the realism of the instrument interactions. [2] uses cosserat models (thin elastic solids) to model a guidewire as a set of straight, non-bendable, uncompressible beams with perfect torque control. Such models consider two types of energy (bending and external contact) and are more accurate and efficient than spring meshes or 3D FEM meshes for thin solids, but real-time simulation is hard to achieve. In [3], the catheter

F. Bello, E. Edwards (Eds.): ISBMS 2008, LNCS 5104, pp. 215–220, 2008.

is modelled as mass-spring particles while the deformable anatomical structure uses FEM. The particles are uniformly distributed along the centreline of the catheter and connected to each other via linear and torsional springs and damping elements.

Our approach, presented in the Section 2, is based on a new hybrid mass-spring system inspired by [2][3] and improved (1) by using the theory of mass spring particles as described in [4] and applied to the catheter or guidewire model, and (2) by using non-bendable springs of equal length to link the particles.

2 Materials and Methods

The goal of our virtual environment is training of the core skills involved in the manipulation of catheters and guidewires in the femoral, iliac and renal arteries, as well as the descending aorta. To achieve this, we offer the possibility of generating virtual vascular networks with varying degrees of complexity, through which a catheter or guidewire can be manipulated. The interaction between the instrument and the 3D vasculature is carried out using the V-Collide collision detection library [5].

The core of the environment is programmed in Visual C++ and interfaced to the H3D library (http://www.h3dapi.org/). This open source library allows taking into account multiple haptic devices and displaying the simulation in 3D. The changes in position of the haptic devices are mapped to the primitive actions that can be performed on the wires, i.e. translation and rotation, and force feedback can be given.

2.1 Synthetic Vasculature Generation

The 3D vasculature integrated in our training environment is composed of synthetic models representing the vessels. Such models are a key component of core skills training since they offer the possibility of practicing the instrument manipulation in vascular trees of varying degrees of complexity.

Fig. 1. Synthetic vascular trees with different parameter values

Our synthetic vasculature generator is able to produce a surface mesh from user specified topologies, where the number of branches and vessel parameters such as length and radius are input via a Python interface. ITK is then used to generate the vessel objects, while surface mesh extraction, smoothing and polygon decimation is performed with VTK. Fig. 1. shows some sample synthetic vessel trees.

2.2 Hybrid Mass-Spring Model

Guidewires and catheters are represented by mass particles separated by non-bendable springs of equal length, λ, as described in [4]. The particle furthest from the insertion point is called x_0, the remaining particles are incrementally labelled from the tip to the body as shown in Fig. 2. Shaped tip particles are designed by specifying the relative position on a local coordinate system. Also shown in Fig. 2 is the bending angle θ, only defined for particles with two neighbouring spring segments.

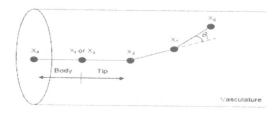

Fig. 2. Particle representation of the guidewire and catheter

With the exception of the particles at each end where a bending force is not defined, all other particles, of index i, are subject to four types of forces given by:

$$
F_i = \begin{cases}
F_{spring} = \sum -k_s (l_{\{i-1,i+1\}} - \lambda)\overline{v}_{\{i-1,i+1\}} \\
F_{bending} = -k_b (\theta - \theta_b)\overline{v}_{sum\{i+1\}} \\
F_{damping} = -k_d \overline{vel} \\
F_{external} = -k_e \overline{n}
\end{cases}
\tag{1}
$$

Where $l_{\{i-1,i+1\}}$ is the distance between the particle's neighbours; $\overline{v}_{\{i-1,i+1\}}$ is the vector between the particle's neighbours; $\overline{v}_{sum\{i+1\}}$ is the direction of the bending force (see Fig. 3), given by v_1+v_2; θ_b is the bias angle (a coefficient used for straightening the instrument's tip); $\overline{vel}$ is the velocity of the particle; $\overline{n}$ is the surface normal; and k_s, k_b, k_d, k_e are the spring, bending, damping and external coefficients, respectively.

Hooke's law of elasticity is applied to the two springs connecting the ith particle to its neighbours. Force is exerted along the spring axis to restore any linear deformation to the rest length. Bending force is associated with the angular orientation between two spring segments as shown in Fig. 3. Bending of the particle is in the $-v_{sum}$ direction. Zero bending force is achieved when two adjacent spring segments are parallel. The external contact force is directed along the face normal of the vessel wall. A damping factor dissipates the amplitude of the springs' oscillations.

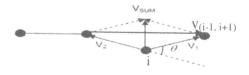

Fig. 3. The bending force of a given particle is in $-v_{sum}$ direction

Since wires are not compressible, we used large spring and damping coefficients. However, this resulted in unstable behaviour of the simulated instrument due to the slow propagation of the user forces. To stabilize the model, we introduce another modification based on [2], assuming that there are non-compressible springs between mass particles and direct propagation of the displacement to all particles in a single integration step. A simple Euler ODE solver is employed for integration.

The instruments can be translated or rotated. Forward translation is implemented in three steps. First, the tip axis is found by subtracting the closest tip particle to the user (x_t on Fig. 2) with the furthest body particle (x_{t+1}). This is the direction of the wire's propagation. Second, each tip particle is displaced by the desired translation along the tip axis. Third, starting from x_{t+1}, each particle is pushed towards the particle in front of it so the springs are at their rest lengths. Reverse translation is done similarly. This implementation, pushing or pulling the tip first, allows reaching equilibrium faster when the instrument is bent and the springs connecting the particles are compressed.

Besides the tip, guidewires and catheters have excellent torque control inside vessels. Ideal torsional control is consequently assumed, as an initial approximation, for their body. Therefore, the position of these particles is not affected when a rotation occurs, i.e. torsional coefficient approaches infinity. Only the tip particles will be affected. When a rotation θ is applied to the instrument, first the tip axis is computed between x_{t+1} and x_t. Then every tip particle rotates around the rotational axis by θ .

2.3 Collision Detection and Collision Response

Collision detection is determined between the instrument particles and the vessel surface mesh. It is checked at every update of the instrument position. Self collision and inter collision between the instrument's particles are not considered.

Within our framework, the collision detection is handled by the open source V-Collide library. It uses regular and axis aligned bounding boxes to perform multi-stage collision detection leading to identification of the triangle collision pair [5].

The thickness of the instruments is not considered here because of the computation time required when the number of vascular mesh nodes is high (which is the case with real vasculature). To keep the implementation simple and efficient, an isosceles triangle is set on each particle circumference. A perpendicular triangle with one vertex facing the positive y-direction is composed. The two triangles effectively create a small 3D collision object around each particle.

Our collision response includes a constant force along the direction of the inner surface normal to prevent vessel perforation. The normal of the triangle where the collision was detected is used to compute the direction of the response force. We assumed for now that the vessels do not deform. Therefore, the response force direction is easier to compute. The force is then applied to the particle in collision to constrain it within the vessel. Because of the springs linking the instrument particles together, the response force is automatically propagated to the other particles.

2.4 Haptic Interface and Feedback

The interventional radiologists manipulate the instruments with one or both hands, translating and rotating them under fluoroscopy. Nevertheless, visualization is not enough to handle the instruments safely for the patient: it is also important to consider

Fig. 4. The VSP Haptic Device from Mentice can track a catheter and a guidewire at the same time and also provides adequate force feedback

the force applied to the instruments and to feel their reactions as they moved through the vasculature. To this end, our virtual environment incorporates a haptic device capable of generating force feedback for both instruments.

The haptic device (VSP) used in this project is a commercial device from Xitact, a division of Mentice. This device (Fig. 4) allows movement tracking (translation and rotation) of two coaxial instruments, such as a guidewire inside a catheter. The VSP also gives force feedback by applying a friction to the instruments. A contrast agent syringe and a balloon inflator are also linked to the VSP but our framework does not yet take these into account.

Our virtual environment uses the VSP to detect instrument motion and update their position on the screen. Depending on the interactions between the instruments and the vasculature, the friction on the instruments is adjusted to provide relevant force feedback. For now, this feedback uses a linear law increasing the friction in function of the length of the catheter/guidewire inserted in the VSP.

3 Tests and Evaluation

Our virtual environment has been tested on synthetic vasculature like the ones presented in Fig.1. An instrument can be placed inside the vasculature and manipulated with the VSP haptic device. The complexity of the vascular models can be progressively increased, and the range and configuration of tools available can be broadened to provide a smooth transition into the complex procedural skills required in everyday practice, therefore providing an efficient training environment.

Fig. 5. Different snapshots of the catheter moving in a 3D synthetic vasculature

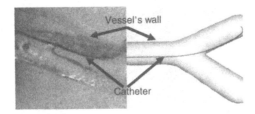

Fig. 6. Comparison of the catheter position in a real environment and the virtual environment

Our environment and the behaviour of the instruments are visually assessed in a realistic setting. Fig. 5 shows three states of the virtual environment where the catheter can be seen in the same vasculature. The results are visually correct as the catheter stays within the boundaries of the vessels and bends in a natural way.

Fig. 6 shows a catheter inside a silicon vascular phantom and a catheter inside a similar simulated bifurcation. The behaviour of the catheter in the physical phantom and simulated vasculature is almost identical.

4 Discussion and Conclusion

The work presented here proposes a new virtual environment for core skills training in Interventional Radiology. Our synthetic vasculature generator can create training scenarios of varying degrees of difficulty. Modelling of catheters and guidewires is based on a novel hybrid mass-spring model, efficient for real-time simulation. Collision detection is performed using V-Collide. A dedicated haptic device is used for tracking real instruments and producing the necessary force feedback. The simulation runs in real time on a PC with a 2.66GHz processor and 2GB of RAM. While this is still work in progress, the level of interest expressed by clinical colleagues is significant and the preliminary results show a realistic behaviour.

Future improvements include the simultaneous manipulation of a second instrument, incorporation of vessel wall behaviour, integration of patient specific vasculature as part of the progression in the training, and further evaluation of our proposed modelling techniques and core skills training methodology.

References

1. Seymour, N.E., Gallagher, A.G., Roman, S.A., OBrien, M.K., Bansal, V.K., Andersen, D.K., Satava, R.M.: Virtual reality training improves operating room performance: results of a randomized, double-blinded study. Ann. Surg. 236, 458–463 (2002)
2. Alderliesten, T., Konings, M.K., Niessen, W.J.: Modeling Friction, Intrinsic of Curvature, and Rotation Guide Wires for Simulation of Minimally Invasive Vascular Interventions. IEEE Transactions on Biomedical Engineering 54(1) (January 2007)
3. Basdogan, C., Ho, C., Srinivasan, M.A.: Virtual Environments for Medical Training: Graphical and Haptic Simulation of Common Bile Duct Exploration. IEEE/ASME Transactions on Mechatronics 6, 269–285 (2001)
4. Witkin, A.: An Introduction to Physically Based Modeling. In: SIGGRAPH 2005 (2005), http://www.cs.cmu.edu/~baraff/pbm/pbm.html
5. Hudson, T., Lin, M., Cohen, J., Gottschalk, S., Manocha, D.: V-COLLIDE: Accelerated Collision Detection for VRML. In: Proc. of VRML 1997 (1997)

Modelling Organ Deformation Using Mass-Springs and Tensional Integrity

Pierre-Frédéric Villard, Wesley Bourne, and Fernando Bello

Biosurgery and Surgical Technology Department, Imperial College London
10th Floor QEQM, St Mary's Hospital
Praed Street, London W2 1NY, UK
{p.villard, f.bello}@imperial.ac.uk

Abstract. In this paper we propose a new framework for deformation modelling based on a combined mass spring and tensional integrity method. The synergy between balanced tension and compression components offered by the tensegrity model helps the deforming organ retain its shape more consistently. We chose the diaphragm as our test object given its heterogeneous composition (muscle and tendon) and its importance in radiotherapy and other interventional procedures. Diaphragm motion can significantly influence surrounding organs such as lungs and liver. Our system permits the control of simulated diaphragm motion and deformation by at least two independent parameters.

Keywords: Deformation model, mass-spring, tensegrity, interventional radiology.

1 Introduction

Our work focuses on the modelling of the intrinsic deformation and motion of organs. Such modelling is highly relevant in the planning and execution of treatments and interventions, as well as for simulation used in the context of training and procedure rehearsal. The vast number of techniques used for organ deformation modelling can be classified into two different categories: the heuristic models such as deformable spline or mass spring, and Continuum-mechanical approach such as the FEM [1]. In the context of procedure training, rehearsal or guidance, real time is vital, therefore we explicitly chose a mass-spring system as it allows reasonably realistic behaviour while supporting near real time performance.

On the whole, organs have an heterogenous composition that can be expected to influence their behaviour and response to external stimuli. In order to incorporate this heterogeneity, we added the tensional integrity formulation (tensegrity) [2] to a mass-spring system to rigidify relevant tissue areas and result in a more realistic behaviour. The diaphragm example presented here illustrates a typical heterogeneity where the muscle tissue can be modelled by a mass-spring system and the tendinous part with tensegrity.

F. Bello, E. Edwards (Eds.): ISBMS 2008, LNCS 5104, pp. 221–226, 2008.

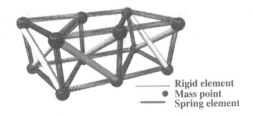

Fig. 1. A tensegrity element (12 mass points, 6 rigid elements and 22 springs)

2   Method

2.1   Tensegrity Definition

The word "tensegrity" comes from a contraction of "Tensional integrity". It is used to define a mechanical system with components that combine tension and compression in such a way as to enable the whole system to receive and apply forces, tensions and pressures. Tension is continuous and compression discontinuous, such that continuous pull is balanced by equivalently discontinuous pushing forces.

Tensegrity systems are composed of two types of elements: the elastic elements, which give the system tension, and the rigid elements, which have a constant length and that will exert compression forces (Fig. 1). This concept of tensegrity is used in many areas. The term "tensegrity" was first explored by artist Kenneth Snelson to produce sculptures such as his 18 meter high Needle Tower in 1968. Imbert [2] states that a wide variety of natural systems, including carbon atoms, water molecules, proteins, viruses, cells, tissues and even humans and other living creatures, are constructed using tensegrity, a common form of architecture. Imbert's view of the cell as a tensegrity structure may help explain why cells in tissue culture spread out and flatten when grown on rigid glass of plastic petri dishes but, when on a flexible surface, the cells contract and become spherical. Therefore, tensegrity is a formulation that seems well suited to organ deformation modelling. The next section explains how tensegrity has been incorporated into our simulation framework.

2.2   Implementation

Our solution makes use of the Java3D API (http://java.sun.com/products/java-media/3D), which is available for all major platforms. We have defined classes to describe topologies, nodes, forces and other required notions to create physical simulations. Topologies can be imported via a mesh parser or can be manually created. Nodes are automatically created along with connections when the parser is used. We have achieved real-time performance by optimising data handling and parameter passing in such a way that the API is notified every time changes have been made to the underlying data. This allows the simulation to be fully interactive and tuneable. The framework takes into account heterogeneity

$$\frac{AA_1}{BB_1} = \frac{B_1B'}{A_1A'} \quad (1)$$

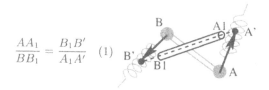

Fig. 2. Rigid links computation after spring influence

to combine object zones with only mass-spring and other zones with tensegrity. Objects are defined by two boundary surfaces. The outward appearance of volume within the two surfaces is created by an initial computation during runtime to create the tensile connections. The algorithm for the main simulation loop is as follows:

1. The elastic forces and the new positions of the nodes are computed as if there were no rigid links using the classical algorithm of the mass-spring method. The node displacements are computed at each time step by a Range-Kutta 2nd order approximation of the fundamental law of the dynamics.

2. The rigid constrains are applied to ensure that the distance between two nodes remains constant as in [3]. Given two nodes A and B linked by a rigid link, A' and B' are their respective positions after applying the classical mass-spring algorithm (as if there was no rigid link). The real positions A_1 and B_1 considering the rigid links are given by Eq. (1) such that A_1 and B_1 remains on the line $A'B'$ (See Fig. 2).

3. The new velocity is computed at each node.

3   Results

Soft tissue deformation and organ modelling are highly complex endeauvors. While a large number of modelling techniques have been proposed with varying degrees of accuracy, the behaviour of certain organs or structures may require the combination of characteristics offered by more than one technique. We propose such hybrid model incorporating the properties of classical mass-spring systems and tensegrity. To evaluate the performance of our hybrid model, we compared its behaviour to that of a simple mass-spring system and used our simulation framework to model and fine-tune diaphragm motion.

3.1   Behaviour of Tensegrity System (TS) and Comparison to Mass-Spring System (MSS)

We used a simple object (cube) to study and compare the behaviour of both systems. The simulations in Fig.3 show the influence of tensegrity. The objects are identical apart from the presence of rigid links. A sinusoidal force is applied to two of the upper nodes in identical directions.

Plotting the displacement for point P in both cubes (Fig. 3.**A**) further illustrates that the tensegrity system more accurately follows the input stimulus

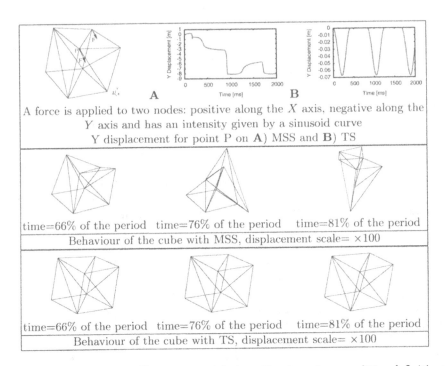

Fig. 3. Cube ($10 \times 10 \times 10m^3$) evolution with time. *Top:* boundary condition definitions, *middle:* results with MSS, *bottom* results with TS.

and is able to handle larger deformations without degradation (maximum Y displacement of $-0.075m$ for the TS and of $-8m$ for the MSS on Fig. 3.B).

3.2 Application to the Diaphragm

Consistent behaviour and ability to withstand large forces are important attributes of our hybrid model. We used these characteristics to model diaphragm behaviour. As mentioned before, the diaphragm is an heterogeneous structure formed of muscular and tendinous tissue, thus ideally suited for our hybrid model. In the case of real diaphragms extracted from manually segmented patients, meshes are created using CGAL (www.cgal.org) library.

A simplified model from a detailed commercial anatomy set (Anatomium 3D by CF Lietzau 3D Special Service) was used to provide the anatomical parts shown in Fig. 4.*left*. Remeshing of the diaphragm was done to obtain a surface mesh composed of 11359 nodes and 22722 triangles. Tendinous tissue and muscle tissue were separated by a plane, and the rigid links of the tensegrity part were created by linking the upper surface to the lower surface with a heuristic algorithm (see Fig.4.*right*).

The boundary conditions as well as the controlling forces are based on a physiological study. First, the anterior-posterior diameter increases because the sternum moves forwards as the ribs are raised. Secondly, the transverse diameter

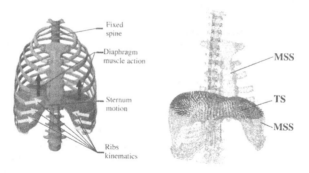

Fig. 4. Real physiological behaviour compared to simulation. Left: physiological motion of diaphragm with the influence of the spine, the sternum, the ribs and its intrinsic muscle contraction/relaxation. Right: wireframe representation of mass-spring elements and rigid elements.

increases due to a "pump handle" movement which elevates the ribs. Both movements can be simulated by the combined motion of the ribs, the sternum and the cartilage. We first assume that the ribs are rigid and their motion can be modelled by a kinematics law based on the finite helical axis method as in [4]. The cartilage-sternum follows the motion of the ribs. This was modelled by a mass-spring system with a very low elasticity in order to ensure that it remains attached to the ribs while having a small internal deformation. The attachment points of the diaphragm are obtained such that the distance to the group ribs, sternum, cartilage is under a given threshold. Then, the vertical dimension increases due to diaphragm movement during inhalation. This motion can be simulated by the muscle action generation borrowed from Thalmann and Nedel [5]. We define and action line as a vertical line (inferior-superior) passing through given points and with a certain radius of action. The forces of contraction during inhalation or relaxation during exhalation are computed such that their intensity varies according to the distance from the action line. The displacement results of this simulation are shown in Fig. 5. To facilitate their visualisation, we display two slices in the anterior-posterior and the axial views.

Fig. 5. Displacement vectors with scale 100 **A:** anterior posterior **B:** axial displacements

4   Discussion

The simple cube experiment illustrates the ability of the tensegrity system to retain the shape of the object, even under considerably larger forces, while the mass-spring cube is crushed under much smaller forces. The tensegrity system more accurately follows the input stimulus and is able to handle larger deformations without degradation.

The diaphragm experiment illustrates the influence of the three phenomena mentioned above: the muscle relaxation that increases the height of the domes, the ribs rotation that presses the diaphragm on each side, and the sternum going backward that pushes the diaphragm. The physiological behaviour of this generic model was discussed with our clinical collaborator, who validated that the anterior-posterior, the transverse and vertical dimensions of the thorax are increased as illustrated by the model.

5   Conclusion

We have presented a model of heterogeneous organ deformation with tissues modelled using both mass-spring and tensegrity system. Comparing the behaviour between the two on a simple object showed the ability of the hybrid model to withstand larger forces and present a more consistent behaviour. Application of this method to diaphragm motion simulation has resulted in a powerful simulation framework that incorporates real boundary conditions, physiological constraints, and allows for the fine tuning of model parameters in real-time as the simulation is running. The different types of diaphragmatic motion are well noticed during the simulation and have been validated by our clinical collaborator. The main contribution of our work is a novel methodology to reproduce heterogeneity by combining elastic and tensile elements.

References

1. Meier, U., et al.: Real-time deformable models for surgery simulation: a survey. Comput. Methods Programs Biomed. 77(3), 183–197 (2005)
2. Ingber, D.: Opposing views on tensegrity as a structural framework for understanding cell mechanics. J. Appl. Physiol. 89, 1663–1678 (2000)
3. Guillaume, A.: 3d simulation of cell biomechanical behaviour. Master's thesis, University Claude Bernard Lyon 1, France (2004)
4. Didier, A.L., Villard, P.-F., Bayle, J.Y., Beuve, M., Shariat, B.: Breathing Thorax Simulation based on Pleura Behaviour and Rib Kinematics. In: IEEE (ed.) Information Visualisation - MediVis, pp. 35–40 (2007)
5. Nedel, L.P., Thalmann, D.: Real time muscle deformations using mass-spring systems. In: CGI, pp. 156–165 (1998)

Author Index

Lecture Notes in Computer Science

Sublibrary 1: Theoretical Computer Science and General Issues

For information about Vols. 1– 4771
please contact your bookseller or Springer

Vol. 4973: E. Marchiori, J.H. Moore (Eds.), Evolutionary Computation, Machine Learning and Data Mining in Bioinformatics. X, 213 pages. 2008.

Vol. 4972: J. van Hemert, C. Cotta (Eds.), Evolutionary Computation in Combinatorial Optimization. XII, 289 pages. 2008.

Vol. 4971: M. O'Neill, L. Vanneschi, S. Gustafson, A.I. Esparcia Alcázar, I. De Falco, A. Della Cioppa, E. Tarantino (Eds.), Genetic Programming. XI, 375 pages. 2008.

Vol. 4967: R. Wyrzykowski, J. Dongarra, K. Karczewski, J. Wasniewski (Eds.), Parallel Processing and Applied Mathematics. XXIII, 1414 pages. 2008.

Vol. 4963: C.R. Ramakrishnan, J. Rehof (Eds.), Tools and Algorithms for the Construction and Analysis of Systems. XVI, 518 pages. 2008.

Vol. 4962: R. Amadio (Ed.), Foundations of Software Science and Computational Structures. XV, 505 pages. 2008.

Vol. 4961: J.L. Fiadeiro, P. Inverardi (Eds.), Fundamental Approaches to Software Engineering. XIII, 430 pages. 2008.

Vol. 4960: S. Drossopoulou (Ed.), Programming Languages and Systems. XIII, 399 pages. 2008.

Vol. 4959: L. Hendren (Ed.), Compiler Construction. XII, 307 pages. 2008.

Vol. 4957: E.S. Laber, C. Bornstein, L.T. Nogueira, L. Faria (Eds.), LATIN 2008: Theoretical Informatics. XVII, 794 pages. 2008.

Vol. 4943: R. Woods, K. Compton, C. Bouganis, P.C. Diniz (Eds.), Reconfigurable Computing: Architectures, Tools and Applications. XIV, 344 pages. 2008.

Vol. 4942: E. Frachtenberg, U. Schwiegelshohn (Eds.), Job Scheduling Strategies for Parallel Processing. VII, 189 pages. 2008.

Vol. 4941: M. Miculan, I. Scagnetto, F. Honsell (Eds.), Types for Proofs and Programs. VII, 203 pages. 2008.

Vol. 4935: B. Chapman, W. Zheng, G.R. Gao, M. Sato, E. Ayguadé, D. Wang (Eds.), A Practical Programming Model for the Multi-Core Era. VI, 208 pages. 2008.

Vol. 4934: U. Brinkschulte, T. Ungerer, C. Hochberger, R.G. Spallek (Eds.), Architecture of Computing Systems – ARCS 2008. XI, 287 pages. 2008.

Vol. 4927: C. Kaklamanis, M. Skutella (Eds.), Approximation and Online Algorithms. X, 289 pages. 2008.

Vol. 4926: N. Monmarché, E.-G. Talbi, P. Collet, M. Schoenauer, E. Lutton (Eds.), Artificial Evolution. XIII, 327 pages. 2008.

Vol. 4921: S.-i. Nakano, M.. S. Rahman (Eds.), WAL-COM: Algorithms and Computation. XII, 241 pages. 2008.

Vol. 4919: A. Gelbukh (Ed.), Computational Linguistics and Intelligent Text Processing. XVIII, 666 pages. 2008.

Vol. 4917: P. Stenström, M. Dubois, M. Katevenis, R. Gupta, T. Ungerer (Eds.), High Performance Embedded Architectures and Compilers. XIII, 400 pages. 2008.

Vol. 4915: A. King (Ed.), Logic-Based Program Synthesis and Transformation. X, 219 pages. 2008.

Vol. 4912: G. Barthe, C. Fournet (Eds.), Trustworthy Global Computing. XI, 401 pages. 2008.

Vol. 4910: V. Geffert, J. Karhumäki, A. Bertoni, B. Preneel, P. Návrat, M. Bieliková (Eds.), SOFSEM 2008: Theory and Practice of Computer Science. XV, 792 pages. 2008.

Vol. 4905: F. Logozzo, D.A. Peled, L.D. Zuck (Eds.), Verification, Model Checking, and Abstract Interpretation. X, 325 pages. 2008.

Vol. 4904: S. Rao, M. Chatterjee, P. Jayanti, C.S.R. Murthy, S.K. Saha (Eds.), Distributed Computing and Networking. XVIII, 588 pages. 2007.

Vol. 4878: E. Tovar, P. Tsigas, H. Fouchal (Eds.), Principles of Distributed Systems. XIII, 457 pages. 2007.

Vol. 4875: S.-H. Hong, T. Nishizeki, W. Quan (Eds.), Graph Drawing. XIII, 402 pages. 2008.

Vol. 4873: S. Aluru, M. Parashar, R. Badrinath, V.K. Prasanna (Eds.), High Performance Computing – HiPC 2007. XXIV, 663 pages. 2007.

Vol. 4863: A. Bonato, F.R.K. Chung (Eds.), Algorithms and Models for the Web-Graph. X, 217 pages. 2007.

Vol. 4860: G. Eleftherakis, P. Kefalas, G. Păun, G. Rozenberg, A. Salomaa (Eds.), Membrane Computing. IX, 453 pages. 2007.

Vol. 4855: V. Arvind, S. Prasad (Eds.), FSTTCS 2007: Foundations of Software Technology and Theoretical Computer Science. XIV, 558 pages. 2007.

Vol. 4854: L. Bougé, M. Forsell, J.L. Träff, A. Streit, W. Ziegler, M. Alexander, S. Childs (Eds.), Euro-Par 2007 Workshops: Parallel Processing. XVII, 236 pages. 2008.

Vol. 4851: S. Boztaş, H.-F.(F.) Lu (Eds.), Applied Algebra, Algebraic Algorithms and Error-Correcting Codes. XII, 368 pages. 2007.

Vol. 4848: M.H. Garzon, H. Yan (Eds.), DNA Computing. XI, 292 pages. 2008.

Vol. 4847: M. Xu, Y. Zhan, J. Cao, Y. Liu (Eds.), Advanced Parallel Processing Technologies. XIX, 767 pages. 2007.

Vol. 4846: I. Cervesato (Ed.), Advances in Computer Science – ASIAN 2007. XI, 313 pages. 2007.

Vol. 4838: T. Masuzawa, S. Tixeuil (Eds.), Stabilization, Safety, and Security of Distributed Systems. XIII, 409 pages. 2007.

Vol. 4835: T. Tokuyama (Ed.), Algorithms and Computation. XVII, 929 pages. 2007.

Vol. 4818: I. Lirkov, S. Margenov, J. Waśniewski (Eds.), Large-Scale Scientific Computing. XIV, 755 pages. 2008.

Vol. 4800: A. Avron, N. Dershowitz, A. Rabinovich (Eds.), Pillars of Computer Science. XXI, 683 pages. 2008.

Vol. 4783: J. Holub, J. Žďárek (Eds.), Implementation and Application of Automata. XIII, 324 pages. 2007.

Vol. 4782: R. Perrott, B.M. Chapman, J. Subhlok, R.F. de Mello, L.T. Yang (Eds.), High Performance Computing and Communications. XIX, 823 pages. 2007.